D1544043

Botanical Miracles

Chemistry of Plants That Changed the World

Botanical Miracles

Chemistry of Plants That Changed the World

Raymond Cooper | Jeffrey John Deakin

CRC Press
Taylor & Francis Group
Boca Raton London New York

CRC Press is an imprint of the
Taylor & Francis Group, an **informa** business

CRC Press
Taylor & Francis Group
6000 Broken Sound Parkway NW, Suite 300
Boca Raton, FL 33487-2742

© 2016 by Taylor & Francis Group, LLC
CRC Press is an imprint of Taylor & Francis Group, an Informa business

No claim to original U.S. Government works

Printed on acid-free paper
Version Date: 20151231

International Standard Book Number-13: 978-1-4987-0428-1 (Hardback)

This book contains information obtained from authentic and highly regarded sources. Reasonable efforts have been made to publish reliable data and information, but the author and publisher cannot assume responsibility for the validity of all materials or the consequences of their use. The authors and publishers have attempted to trace the copyright holders of all material reproduced in this publication and apologize to copyright holders if permission to publish in this form has not been obtained. If any copyright material has not been acknowledged please write and let us know so we may rectify in any future reprint.

Except as permitted under U.S. Copyright Law, no part of this book may be reprinted, reproduced, transmitted, or utilized in any form by any electronic, mechanical, or other means, now known or hereafter invented, including photocopying, microfilming, and recording, or in any information storage or retrieval system, without written permission from the publishers.

For permission to photocopy or use material electronically from this work, please access www.copyright.com (http://www.copyright.com/) or contact the Copyright Clearance Center, Inc. (CCC), 222 Rosewood Drive, Danvers, MA 01923, 978-750-8400. CCC is a not-for-profit organization that provides licenses and registration for a variety of users. For organizations that have been granted a photocopy license by the CCC, a separate system of payment has been arranged.

Trademark Notice: Product or corporate names may be trademarks or registered trademarks, and are used only for identification and explanation without intent to infringe.

Library of Congress Cataloging-in-Publication Data

Names: Cooper, Raymond, 1949- | Deakin, Jeffrey John.
Title: Botanical miracles : chemistry of plants that changed the world /
Raymond Cooper and Jeffrey John Deakin.
Description: Boca Raton : Taylor & Francis, 2016. | "A CRC title." | Includes
bibliographical references and index.
Identifiers: LCCN 2015048269 | ISBN 9781498704281 (alk. paper)
Subjects: LCSH: Medicinal plants. | Botanical chemistry. | Chemistry, Organic.
Classification: LCC RS164 .C724 2016 | DDC 615.3/21--dc23
LC record available at http://lccn.loc.gov/2015048269

Visit the Taylor & Francis Web site at
http://www.taylorandfrancis.com

and the CRC Press Web site at
http://www.crcpress.com

To

John B. Stuffins

An inspirational teacher and a lifelong friend to us both

Contents

Foreword xv
Acknowledgments xvii
Authors xix

Chapter 1 Introduction 1

Aims and Purpose 1
Importance and Role of Natural Products 1
Organic Chemistry 5
A Note on the Scope 7
A Note on Nomenclature 7
A Note on Botanical Names 7
A Note on References 7

Chapter 2 Medical Marvels 11

Introduction 11
Ancient Medicine 11
Central America's Humble Potato! 13
Steroids 13
Ancient Forms of Contraception 13
The Search for the Modern Birth Control Pill: The Hormones, Estrogen and Progesterone 14
Isolation of Natural Diosgenin from the Mexican Yam 15
Chemical Magic in the Laboratory: Synthesis of Norethindrone 16
Biological Studies of Progesterone to Prevent Ovulation 17
Medical Approval and Social Acceptance 17
The Profound Social, Cultural and Economic Impacts of Reliable, Oral Contraception 18
Transformation: Global Emergence of the Yam 18
Hydrocarbons 19
Saturated Hydrocarbons 19
Alkanes 19
Cycloalkanes 19
Crude Oil and Industrial Fractional Distillation 20
Benzene and Aromatic Compounds: Physical and Chemical Properties 22
Theory of the Molecular Structure of Benzene 23
Summary of the Characteristics of Aromatic Compounds 24
Europe Solves a Headache! Emergence of Aspirin 26
Salicin and Salicyclic Acid 26
Carboxylic Acids 26
Phenol and Phenolic Compounds 27

Acetylation of the Hydroxyl Group of Salicylic Acid to Form Aspirin ... 28
Attacking Malaria: A South American Treasure (but Not Gold) and a Chinese Miracle ... 31
Malaria ... 31
Cinchona ... 32
Isolation of Quinine ... 32
Synthesis of Quinine in the Laboratory ... 33
Uses of Quinine ... 33
Artemisinin ... 34
Carbon–Oxygen Single Bonds and Oxygen–Oxygen Single Bonds in Organic Compounds ... 34
The Peroxide Link and the Peroxide Bridge in Artemisinin ... 36
A Steroid in Your Garden ... 38
The Foxglove and Digoxigenin ... 38
Esters Are Formed from Alcohols and Acids ... 40
Properties and Uses of Esters ... 41
Introducing Cyclic Esters Known as Lactones ... 42
Lactones as Building Blocks in Nature ... 42
Vitamins ... 42
Vitamin C, Ascorbic Acid ... 43
Commercial Uses of Lactones ... 44
Africa's Gift to the World ... 46
Discovery of the Periwinkle Plant and Its Properties ... 46
Serendipity ... 47
Angela's Story ... 48
Modern Research and Therapeutic Value ... 48
The Alkaloids, Vincristine and Vinblastine ... 49
Isolation of Vincristine and Vinblastine from Plants ... 50
Fractional Distillation ... 50
Acid–Base Extraction ... 50
Saving the Pacific Yew Tree ... 52
The Nature of Paclitaxel ... 52
Medical Value of Paclitaxel ... 53
Modern-Day Preparation of Paclitaxel ... 53
Terpenes and Isoprene as a Building Block in Nature ... 53
Isomers ... 54
Isomers and Stereochemistry ... 54
Isomers and Chirality ... 55
Nuclear Magnetic Resonance (NMR) Spectroscopy and Molecular Structure ... 56

Chapter 3 Modern Miracles of Foods and Ancient Grains ... 59

Introduction ... 59
Rediscovering Traditional Grains of the Americas: Chia and Quinoa ... 60

Chia ... 60
Quinoa ... 61
Gluten Sensitivity ... 61
Lipids ... 62
Fats, Oils and Human Health ... 63
Waxes ... 64
Vegetable Oil and Biodiesel ... 64
Foods of the Fertile Crescent: Ancient Wheat ... 66
Possible Origins of Ancient Wheat ... 66
Crop Domestication ... 67
Crop Germination ... 67
Phenylpropanoids ... 68
Lignans ... 68
Lignin ... 69
Major Cultivated Species of Wheat ... 69
Modern Wheat ... 70
Wheat Genetics ... 71
Nutritional Importance of Wheat and Grinding ... 71
Importance of Proteins ... 72
Peptide Bond ... 73
The Condensation Reaction ... 73
Proteins and Amino Acids ... 74
The Molecular Structure of Proteins and NMR ... 75
Fascinating Influence of Chirality on Life Systems on Earth ... 76
The Shikimic Acid Pathway ... 76
Eugenol and Rosmarinic Acid ... 78
Society's Challenges in Enhancing Agricultural Production: Crop Yield and "Green Issues" Including Genetic Modification of Crops ... 78
Asian Staple: Rice ... 82
Carbohydrates and Saccharides: Oxygen in the Organic Ring ... 82
Monosaccharides ... 83
Disaccharides ... 83
The Glycoside Link ... 84
Oligosaccharides ... 85
Polysaccharides ... 85
Chemical Properties of Carbohydrates: Monosaccharides ... 86
The Value of Monosaccharides, Disaccharides and Polysaccharides to Living Organisms ... 87
Human Nutrition ... 87
Commercial Uses of Carbohydrates ... 87
Brewing of Beer ... 87
Adhesives and Stiffening Agents in Fabrics ... 88
Bioplastic ... 88
Explosives ... 89
Chinese Cordyceps: Winter Worm, Summer Grass ... 90

The Life Cycle of *Cordyceps sinensis* 90
The Perceived Health Benefits of *Cordyceps sinensis* 90
What Is Fermentation? 93
Chemistry of Fermentation: Redox Reactions 93
Industrial Production of Bioethanol 94
Garlic and Pungent Smells 96
Garlic 96
Organosulfur Compounds 97
Amino Acids Which Contain Sulfur 97
Fossil Fuels and Air Pollution 99

Chapter 4 Beverages 101

Introduction 101
Tea: From Legend to Healthy Obsession! 102
The Growing and Processing of Tea 104
Tea as a Modern Medicine 105
Green Tea Helps against Cancer 106
Catechins: Key Phenolic Constituents in Green Tea 106
The Properties of Phenol and Phenols 107
Phenols Compared with Alcohols 107
Electrophilic Substitution Reactions of Phenol 108
Interaction with Light in the Visible Part of the Spectrum of Electromagnetic Radiation 108
Cocoa (Cacao): Food of the Gods 110
Origins 110
Cacao and the Aztecs 110
Processing Cacao Beans 111
Cocoa and Cardiac Health 112
Cocoa and Diabetes 112
Chemical Constituents of Cocoa 113
Free Radicals and Antioxidants 113
Flavonoids (or Polyphenols) 113
Coffee: Wake Up and Smell the Aroma! 115
Early Use 116
Coffee and Caffeine 117
Zwitterions 118
Cyclic Aromatic Amines 118
The Process of Decaffeination of Coffee 119
Maca from the High Andes in South America 122
The Maca Plant 122
Maca as a Beverage and a Food 122
The Medicinal Value of Maca 123
Chemical Composition of Maca 123
Indole 124
Key Chemical Properties of Indole 124

Electrophilic Substitution ... 124
Indole and Bases ... 125
Acid–Base Extraction ... 125
Purification of Indole Alkaloids by Acid–Base Extraction ... 125
Indole Alkaloids ... 126
Strychnine ... 126
Ergotamine ... 127
Lysergic Acid Diethylamide (LSD) ... 127
Curare ... 128

Chapter 5 Euphorics ... 131

Introduction ... 131
Morphine: A Two-Edged Sword ... 132
Joe and Mike: The Two-Edged Sword ... 132
Production of Morphine in the Natural World ... 134
Purification, Chemical Composition and Properties of Morphine ... 135
Chemical Composition and Properties of Codeine ... 136
Chemical Composition and Properties of Heroin ... 136
Medicinal Uses of Morphine ... 138
Summary ... 138
Cannabis and Marijuana ... 140
Cannabis ... 140
Introducing Cannabinoids ... 140
Terpenes ... 141
Classification of Terpenes ... 142
Terpenes and Elimination Reactions ... 142
The Condensation Reaction ... 143
Coca and Cocaine ... 145
Coca and the Coca Plant ... 145
Cocaine ... 145
Chemical Properties of Cocaine ... 146
Isolation of Cocaine ... 147
Legitimate Applications of Cocaine ... 147
Tobacco: A Profound Impact on the World ... 149
Tobacco ... 149
Nicotine ... 151
Amines ... 151
Heterocyclic Aromatic Compounds ... 151
Pyridine ... 152
Pyrrole ... 152
Thiophen ... 152
Amides ... 153
Properties of Nicotine ... 154
Uses of Nicotine ... 154

Agriculture.......154
Human Health and Pharmacy.......155

Chapter 6 Exotic Potions, Lotions and Oils.......157

Introduction.......157
A Plant from the East Indies: Camphor.......158
The Classification and Structure of Camphor.......158
Diverse Uses of Camphor.......158
Extraction of Camphor by Steam Distillation.......159
The Carbonyl Group and Nucleophilic Addition Reactions.......160
Infrared Spectroscopy and the Determination of the Structure of Organic Molecules.......161
Absorption of Infrared Radiation: The Greenhouse Effect and Global Warming.......161
Biblical Resins: Frankincense and Myrrh.......165
Sources and Uses of Frankincense and Myrrh.......165
Frankincense.......166
Myrrh.......168
Classification of Cyclic Terpenes.......168
Steric Hindrance.......171
European Lavender.......173
European Lavender.......173
Distribution.......174
Value of Lavender.......175
Food and Drink.......175
Medicinal Applications.......175
Cosmetics.......175
Soap and Detergent.......175
Essential Oils: Hydrosol from Lavender.......175
Hydrosols and Colloids.......176
Vegetable Oils and Fats.......176
Glycerol.......177
Triglycerides.......177
Hydrolysis of Esters.......177
Soap.......177
How Soap Works.......178
Global Aloe.......180
Aloe vera.......180
Aloe vera and Skin Treatment.......181
Aloe vera and Polysaccharides.......181
Galactomannans.......181
Hydraulic "Fracking".......182
Anthraquinones Found in *Aloe vera*.......183
Aloin.......183
Emodin.......184

Chromatography ... 185
Thin-Layer and Column Chromatography ... 185
Gas Chromatography and Gas–Liquid Chromatography ... 186
High-Pressure Liquid Chromatography ... 186
Ion-Exchange Chromatography ... 186

Chapter 7 Colorful Chemistry: A Natural Palette of Plant Dyes and Pigments ... 189

Introduction ... 189
Our World of Green Plants: Human Survival ... 191
Green Plants ... 192
Flowering Plants and Conifers ... 192
Algae ... 192
Lichens ... 192
The Vital Process of Photosynthesis ... 193
Chlorophyll ... 193
Chlorophyll and Color ... 195
Chromophores ... 195
Molecular Interaction with Electromagnetic Radiation ... 197
Ultraviolet Absorption Spectroscopy and Organic Chemistry ... 198
Ultraviolet Absorption Spectroscopy and Molecular Structure ... 199
Green Plants and Limiting Climate Change ... 200
Saffron and Carotenoids: Yellow and Orange Dyes ... 204
Saffron: The Plant ... 204
Crocin: The Extract from Saffron ... 205
Carotenoids and Autumnal Colors ... 205
Carotenoids, Carotenes and Xanthophylls ... 207
Carotenes ... 207
β-Carotene ... 207
Lycopene ... 207
Isolation and Extraction ... 208
Xanthophylls ... 208
Crocin ... 208
Lutein ... 208
The Allyl Functional Group ... 209
Human Health: Vitamin A, Retinol and Retinal ... 210
Woad (*Isatis tinctoria*) and Indigo ... 212
Indigo through the Ages ... 212
Indigo from Plants ... 213
Extraction of Indigo ... 214
Physical and Chemical Properties of Indigo ... 214
Indigo and Dyeing Textiles ... 215
Levi Strauss and Wrangler Jeans ... 215
"Lincoln Green" Worn by Robin Hood ... 215
Azo Dyes ... 217

Red Dyes from Henna, Dyer's Bugloss and Madder........ 220
Quinones as Building Blocks in Nature........ 220
Naphthoquinones........ 221
Lawsone, a Red-Orange Dye from Henna........ 221
Alkannin, a Red-Brown Dye from Dyer's Bugloss........ 222
Anthraquinones........ 222
Alizarin Red from Madder Root........ 223
Parietin from the Lichen, *Xanthoria parietina*........ 223
Textile Dyes: Colorfastness and Mordants........ 225
Solubility of Dyes in Water........ 225
Colorfastness........ 225
Mordants........ 225
Reversible Colors in Flowers, Berries and Fruit........ 228
The Inestimable Value of Color in Flowers, Berries and Fruit... 228
The Beautiful Color of Autumn Leaves........ 229
Flavonoids........ 229
Flavonoids and Anthocyanins........ 230
Anthocyanins: Reversible Dyes........ 231
Reversible Dyes and Acid/Base Indicators........ 231
Litmus........ 231
Methyl Red and Methyl Orange........ 232
Phenolphthalein........ 233
Phytochrome: A Reversible Pigment and a Biological Light Switch: Photoperiodism........ 233

Glossary........ 237

Index........ 243

Foreword

The importance of natural products and their impact on human development and society cannot be overstated. Natural products can be interpreted as pure chemicals or their mixtures from nature or even the crude materials afforded by Mother Nature herself, for example, timber, resins and crude herbal drugs. The sources of these chemicals are manifold with key examples from insects, plants, microbes and terrestrial and marine animals. To give an indication of the value of these materials, almost 25% of all of our prescription drugs owe their origins to a natural source, and in the area of antibiotics and anticancer agents this figure is nearer to 40%. It is hard to imagine a world where natural products do not play a pivotal role in all areas of human endeavors.

Our anticancer drugs are a classic case of this impact, with drugs such as Paclitaxel from the Pacific Yew tree (*Taxus brevifolia*) for ovarian and breast cancer, and Trabectedin from a marine tunicate (*Ecteinascidia turbinata*) for soft-tissue cancers. Both of these drugs are superb examples from our global chemical diversity. Their complex shapes, beautiful functionality and exquisite biological activity exemplify the value of chemicals from natural sources. Even our cholesterol-lowering drugs owe much to natural chemicals; with lovastatin from the filamentous fungus *Aspergillus terreus* being a template for our modern statin drugs, widely used globally to lower the risk of heart disease.

Natural products not only have importance for the science of single chemical entity drugs, but for their impact on crude drugs and complex mixtures. An understanding of the chemical complexity of crude drugs in systems of medicine (e.g., traditional Chinese medicine) is vital to ensure quality, stability and efficacy in formulations and preparations. This can be extended to studying the ethnobotanical uses of medicines, to understand how traditional healers use these substances, which can lead to new drugs and equitable benefit sharing amongst communities. The goal of this process should be the preservation of traditional knowledge for future generations.

Our understanding of natural products has also contributed to our appreciation of our environment; nature can be "nasty," and many of the most potent toxins are naturally produced and, where obvious, are historically avoided, for example, Hemlock of Socrates fame. These substances are of varying degrees of severity ranging from contact poisons (*Heracleum* sp.), to inedible toxic berries (*Atropa belladonna*) through to deadly mushrooms (*Amanita phalloides*). An awareness of their taxonomy, phytochemistry and biological activity has resulted in a greater understanding of their impact.

As an extension of this, our ability to understand the effects of drugs on human consciousness is also related to natural product science. The human desire for intoxication through alcohol and drugs of abuse such as heroin has enabled us to appreciate their damaging effects on society, their cultural usage and even their augmentation to medicinal properties. Examples of both sides of this can be seen with Cannabis (*Cannabis sativa*), a drug now legal in several states of the United States, and having a European product licence for ameliorating the pain and spasm of patients with multiple sclerosis. Unfortunately, such drugs are often accompanied with a darker side, for example, tetrahydrocannabino, the active compound within cannabis, has led to the development of many synthetic cannabinoids, the so-called "legal highs" or novel psychoactive substances. These materials have exceptionally high potency at the CB_1 receptor, and may well have long-term harms leading to psychosis and schizophrenia.

Natural products also have a profound impact on plant reproduction: Many insects are able to detect colored compounds, ultraviolet-active compounds and volatile attractants and, therefore, pollinate many of our crop plants as they are attracted to the plant flowers. This has vital importance for the development of ecosystems and for mammalian life. We are often oblivious to the fact that natural products have made enormous contributions to human healthcare, improved our understanding of our interactions with nature and made huge impact on the development of human society and other species on our planet.

Botanical Miracles: Chemistry of Plants That Changed the World is, therefore, aptly named as it beautifully makes the case for how natural products have impacted on our lives, culture and our development. Dr. Raymond Cooper and Dr. Jeffrey Deakin have written a text aimed at encouraging young people at the "A"-level/high school to early undergraduate studies of the value and utility of plant natural products. This text builds on Dr. Cooper's earlier text *Natural Products Chemistry: Sources, Separations and Structures* published in 2014, which highlights the basic science behind natural product science. *Botanical Miracles: Chemistry of Plants That Changed the World* takes the reader through the medicinal impact and classes of the main phytochemicals, how our food and drink has transformed our society, the usage of psychoactive materials and how plant natural products are used as *cosmeceutical* preparations. Finally, the chemistry of color is covered, ranging from the chlorophylls and photoaccessory pigments through to the pigments that shaped commercial colorants in the industrial revolution, through to the natural colors of berries in our foodstuffs that profoundly impact on our health.

This book has great utility and has been written with a sense of enthusiasm that will be infectious to the reader. It will find an avid audience amongst those who wish to understand the enormous contributions that plants have made to our world, and the further opportunities that exist from this and other natural sources. Drs. Cooper and Deakin have put together a fine and highly readable text, and it is with great confidence that this lovely book can be recommended to those who wish to further appreciate the magnificent contributions of plants and natural products to our world, our health, security and comfort.

Professor Simon Gibbons
London, UK

Acknowledgments

The authors wish to acknowledge the encouragement, unfailing support and optimism of Dr. Gaynor Sharp from the Association for Science Education in the United Kingdom.

We offer thanks for the guidance and patient help of Hilary LaFoe and Joselyn Banks-Kyle at Taylor & Francis/CRC Press, and for the helpful suggestions and editing of Dr. Vivien Shanson, formerly at Imperial College, University of London, UK.

We appreciate access to the excellent facilities of The Wellcome Library in London and The Peter Raven Library in the Missouri Botanical Garden, St Louis, Missouri.

This book would not have been possible without the brilliance of the many natural-products chemists who have dedicated their research over many decades to the challenge of elucidating the chemical structures and properties of the exquisite and complex molecules found in the world of plants. They leave a lasting legacy.

While their names are too numerous to mention, we wish to recognize especially the inspirational leadership of the late Professor Carl Djerassi who passed away while we were writing this book. He made many valuable contributions in the field of natural products chemistry and, in particular, to the development of the contraceptive pill from the Mexican yam. We trust we have represented his contributions accurately.

Authors

Raymond (Ray) Cooper was born in the United Kingdom, earned a BSc in chemistry at the University of London and a PhD in the chemistry of natural products from the Weizmann Institute in Israel. His dissertation researched the ancient wild wheats of the Middle East, examining their germinating properties and chemical profiles.

After completing a postdoctoral fellowship at Columbia University, New York, Ray spent 15 years in discovery research of plant natural products in the pharmaceutical and biotechnology industries. He then moved to the nutraceutical and dietary supplements industry to develop botanicals from traditional Chinese medicine including ginkgo, cordyceps, red yeast rice, green tea and many other botanical medicines.

Currently, Ray is a visiting professor at the Hong Kong Polytechnic University, Hong Kong. He is a Fellow, Royal Society of Chemistry, UK and honorary visiting professor at the College of Pharmacy, University of London, UK. Ray is also an associate of the Missouri Botanical Gardens, St Louis, Missouri.

He is a member of the American Pharmacognosy Society, having published over 100 research papers and received the 2014 Varro Tyler Award for Contributions to Botanical Research.

He is an associate editor, *Journal of Alternative & Complementary Medicine*, an editor of five books and coauthor of *Natural Products Chemistry: Sources, Separations and Structures*, published in 2014 by Taylor & Francis.

Jeffrey (Jeff) John Deakin earned a first class honors degree in chemistry from the University of London followed by a doctor of philosophy degree in physical chemistry from the University of Cambridge.

As a teacher of science, Jeff was head of the chemistry and physics departments in grammar and comprehensive schools in the United Kingdom. Later, he became deputy director of education at Lincolnshire County Council responsible for county education services for pupils in some 400 schools. Jeff then went on to establish and direct his own education services company through which he acted as an advisor to secondary schools, local education authorities, HM Government education agencies and the Department for Education.

Jeff was a founding member and nonexecutive director of a multiacademy educational trust, formally approved by the Department for Education, which aims to secure and sustain school improvement by providing leadership and support, by working with governing bodies to strengthen their leadership and strategic delivery and through contracted work with school leaders and their teams. At the same time he was also the chairman of the governing body of one of the largest academies in the secondary sector of education within the United Kingdom.

Jeff is a Fellow, Royal Society of Chemistry, UK and a member of the Curriculum and Assessment Working Group at the Royal Society of Chemistry in London, which is reviewing the national curriculum in chemistry in each of the four home nations of the United Kingdom of Great Britain and Northern Ireland.

1 Introduction

AIMS AND PURPOSE

Botanical Miracles: Chemistry of Plants That Changed the World is a platform to present an educational journey and to supplement and extend the teaching curriculum by providing context for learning in organic chemistry. The book is intended to inspire, enhance and enrich an inquiring mind through a multidisciplinary approach; embracing science, medicine, the natural environment, geography and history (Figure 1.1). Through examples of well-known plants and modern scientific miracles, the book tells the stories of key natural products and their extracts with emphasis upon sources, applications and science. These extracts are today associated with important drugs, nutrition products, beverages, perfumes, cosmetics and pigments. The book is written for those chemistry students who are in the upper age groups at high school or at early university level.

When used in the broader context of general studies classes, extracts selected by tutors from a wealth of cross-curriculum material may stimulate informed debate about the relationship between scientific development and commercial exploitation of the products of plant biochemistry fostering exploration of related business, ethical and social issues.

Furthermore, the material is also intended to excite the interest of scientifically literate people who wish to broaden their horizons on the basis of personal or professional motivation.

IMPORTANCE AND ROLE OF NATURAL PRODUCTS

Plants have changed the world!

Botanical Miracles: Chemistry of Plants That Changed the World traces the stories of wonderful plants which have shaped civilizations, trade and conquest. In order to create this book we have chosen special plants with unique stories—exquisite plants providing medicines, foods, beverages, perfumes and pigments. Particular plants have been used by man for millennia and form part of ancient folklore. Beneficial plants have been selected through a combination of trial and error, as well as speculation and intuition.

Medicinal plants are an important part of human history, culture and tradition. Plants have been used for medicinal purposes for thousands of years. Wild animals instinctively know which plants to consume (and avoid). Birds build their nests using plants, which are bacteria resistant in order to protect their young. Early humans knew of a plant's healing capabilities. Cave paintings dating as far back as 13,000–25,000 BC have been discovered depicting the use of plants as medicine.

Anecdotal and traditional wisdom concerning the use of botanical compounds is documented in the rich histories of traditional Chinese medicine and Ayurvedic

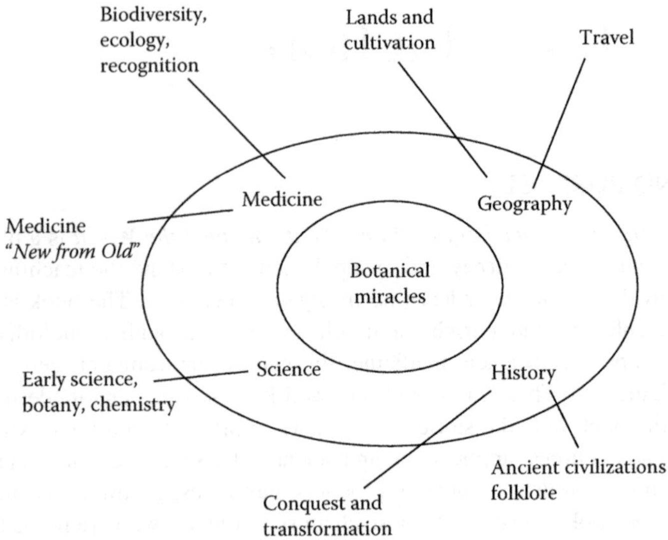

FIGURE 1.1 Multidisciplinary: inspiring, enjoying, and embracing science.

medicine in India, and the treatments administered by Native Americans. Many medicinal plants, spices and perfumes changed the world through their impact on civilization, trade and conquest. In the West, people have successfully and ingeniously adapted indigenous, early knowledge for application in modern culture and medicine. An intention in the book is to trace a route through history from ancient civilizations to the modern day showing the important value of natural products in medicines, in foods and in many other ways.

Over the years, plants, which were neither bitter nor poisonous were selected. People learned through experience to select those plants that offered nutritional value and acceptable taste or could be used in cooking, ground up as a spice, or used as a tea. Through empirical trial and error, beneficial plants with medicinal properties were selected leading to life-saving medicines.

Traditional medicine often aims to restore balance to the body. By contrast, Western allopathic medicine involves a well-defined, single chemical entity with specific medicinal properties.

Many of our so-called "modern day" drugs have origins in ancient medicine. There has also been a significant increase in awareness of and use of natural "alternative" medicines. A renaissance in the appreciation of medicinal plants is underway.

Folk medicine is commonly characterized by the application of simple indigenous remedies. People who use traditional remedies may not understand, in our terms, the scientific rationale for why they work, but know from personal experience that some plants can be highly effective.

Study of chemicals from nature as sources of drugs is called pharmacognosy, derived from two Greek words, "pharmakon," or drug, and, "gnosis," or knowledge.

Its scope includes the study of the physical, chemical and biological properties of existing drugs as well as the search for potential drugs from natural sources.

The field of ethnobotany has also expanded. Ethnobotany is the scientific study of the relationships, which exist between people and plants. The discipline requires a variety of skills: botany—the identification and preservation of plant specimens; anthropology—understanding of the cultural concepts around the use of plants; and chemistry—understanding the behavior and chemical properties of natural products.

Amazon rainforest tribes have lived for millennia in the lush, tropical rainforest subsisting by hunting and gathering. They have learned the medicinal value of plants and animal species through their shamans and medicine men, and have accumulated empirical knowledge irreplaceable today by modern science. One famous example is the use of curare. The word "curare" comes from the South American Indian name for arrow poison which is *ourare*. It is obtained from the bark of the South American plant, *Chondrodendron tomentosum*. Among all Amazon rainforest plants, curare has been invaluable for a traditional hunting technique: blowgun hunting (Figure 1.2).

These tribesmen craft hunting darts, about 25 cm long, which are made from the stalks of palm leaves, smoothed and sharpened using piranha teeth. A dart tipped with curare (Figure 1.3) can kill a bird in a few seconds, a man in 5 minutes or a deer in 30 minutes. Curare is a fast-acting poison. It is not used to kill the prey but to cause paralysis. Death comes by asphyxia when the lungs of the prey are paralyzed. One of the active agents of curare is now known to be tubocurarine, a toxic alkaloid and skeletal muscle relaxant. Intriguingly, Amazonian rainforest tribes are able to eat their prey subsequently without any untoward effects.

FIGURE 1.2 **(See color insert.)** Yagua tribesmen in Iquitos, Peru can kill a monkey 30 m away with a blowgun. (With permission under terms of GNU Free Documentation License.)

FIGURE 1.3 Blowgun darts are used in the Peruvian Amazon, tipped with a deadly poison from one of the Amazonian rainforest plants (curare). (With permission under terms of GNU Free Documentation License.)

Important examples of the discovery of valuable medicinal plants, which are of inestimable benefit to mankind are presented in Chapter 2. It might be said that many people in the world only know of indigenous medicines based upon ancient tradition, yet we in the West are recipients of these wonderful gifts without perhaps ever appreciating their origins.

This book provides many examples of how ancient civilizations survived, thrived and flourished by using local plants. Through trade, conquest, vision and ingenuity, the West successfully absorbed and then developed indigenous knowledge into its culture and economy. In more recent times, chemical examination of these extracts has led to the identification of many natural product constituents leading to modern foods and beverages of considerable value to humankind which are described in Chapters 3 and 4 of this book. A significant example was the use of lime fruits by the British navy which overcame scurvy even though it took another 200 years to discover that the reason was the presence of vitamin C in the citrus.

In the nineteenth century, substances were identified as poisons or those which could affect mood—several are described in detail in Chapters 5 and 6. Two powerful compounds, morphine and strychnine, were isolated from different plants in 1815 and 1819, respectively, although their actual chemical structures were not defined for another 100 years! Numerous examples of pharmaceutical and over-the-counter drugs were derived directly from compounds found in nature, or indirectly through chemical modification of the basic chemical structure.

In Chapter 7 of this book, we celebrate the vital importance of all the plants of the natural world. Directly and indirectly, humankind and other members of the

Animal Kingdom are wholly dependent upon plants. The Plant Kingdom is dominated by green plants converting carbon dioxide and water into a vast and diverse array of organic compounds, providing vital resources in food, clothing and shelter, and releasing indispensable oxygen into the atmosphere for respiration.

We appreciate the beauty of color in the natural environment, which inevitably influences art and science through the exploitation of natural pigments. Therefore, in Chapter 7, we also highlight the colorful chemistry of natural products. Have you ever wondered what makes the world of plants so colorful? Why are the leaves green and then turn to wonderful hues of yellow and red in the autumn? The explanation is due to the presence of natural pigments in the plants. Indeed, historically, many pigments from plants were used as dyes, and these were revered also for possessing "magical properties" with the power to heal and to keep evil spirits at bay. Mystery and superstition surrounded the extraction of the essence. An example is provided by *Woad*. This is a plant, *Isatis tinctoria*, which yields the purple-blue color, indigo, scarce in the natural world. The desire and need for different colored dyes stimulated much research by chemists in the nineteenth century, which contributed to transformation of society from its historic agrarian foundation to the modern industrial era.

ORGANIC CHEMISTRY

Having a valency of four, a carbon atom may be bound covalently to another carbon atom through a single, a double or a triple bond. Significantly, carbon atoms can bond together to form different structures, such as chains and rings. Carbon can also combine covalently with other elements, typically hydrogen, oxygen, nitrogen, sulfur and phosphorus. As a result, infinite combinations and permutations are possible, which lead to a bewildering array of naturally produced organic molecules: some simple and others complex.

Despite this complexity, many metabolic pathways in plants and animals begin from simple building blocks. One of the most famous metabolic cycles owes its existence to Krebs, who worked out the pathway of simple primary metabolites (Figure 1.4). Metabolites are intermediates in metabolic processes in nature and are usually small molecules; examples are ethanol, acetic acid, citric acid and lactic acid, or cell constituents such as lipids and vitamins. A primary metabolite is involved in the growth, development and reproduction of an organism.

Whilst secondary metabolites are not directly involved in these basic, vital processes, they have functions, which are still of great value to the organism in providing antibiotic properties or pigments for camouflage, for example.

Although the chemistry and structure of many substances from nature is complex, the natural products introduced in this book are scientifically fascinating and have a rich history.

Natural products are described in relation to the basis of the organic chemistry curricula for early university courses in the United States, and for GCE "A" level and early university courses in the United Kingdom. The aims are to inspire, inform and extend students while helping them gain an appreciation of these complex molecules.

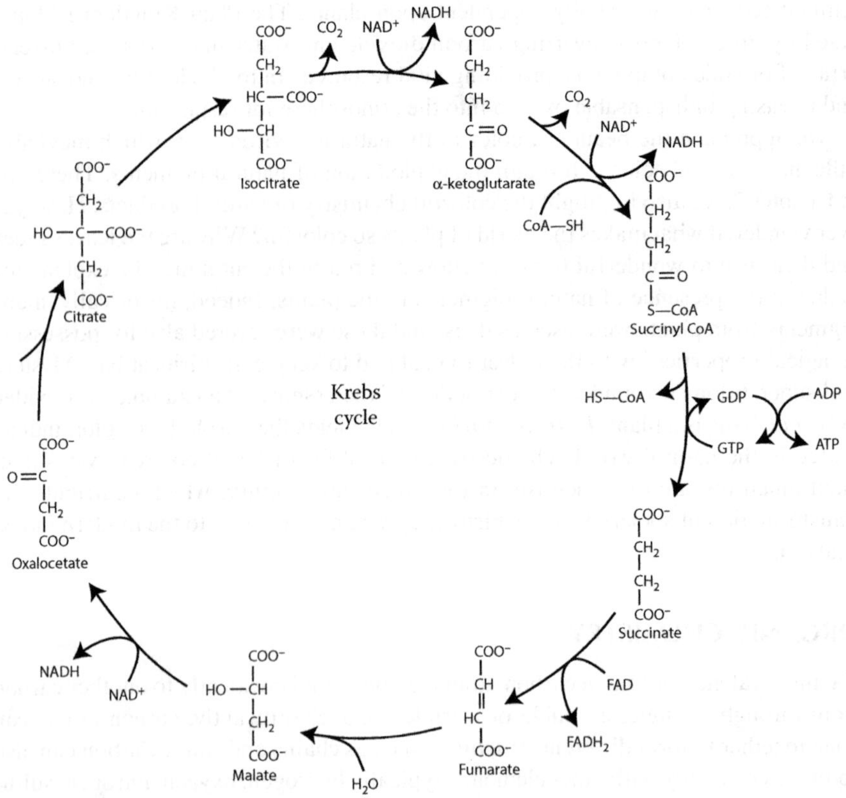

FIGURE 1.4 Krebs cycle, illustrating that metabolic processes and enzymatic reactions begin from simple building blocks. (From http://www.uic.edu/classes/bios/bios100/lecturesf04am/lect12.htm)

Although it is not the intention to cover either the curriculum or preparation methods comprehensively as an instructional textbook might, study of an extensive range of functional groups and concepts is interwoven in these chapters:

- Covalent bonding
- Alkanes
- Halo alkanes
- Cycloalkanes
- Compounds containing the carbonyl group
- Alcohols
- Esters
- Alkenes
- Alkynes
- Aromatic chemistry
- Amines and amides

- Amino acids (including the peptide link and proteins)
- Phenols
- Heterocyclic compounds
- Polymers
- Isomerism
- Chirality.

We draw on the chemistry of these functional groups to show how they influence the chemical behavior of the building blocks, which make up large and complex molecules.

Isomers and polymers are examined, since stereochemistry subtly and profoundly influences the interactions of large, complex molecules found in nature, and hence, their physical and chemical properties.

There is an introduction to techniques, which are used to investigate the structure of large molecules, and to the application of chromatography in purification and analysis.

For the instructor, the text supplements the core curriculum and offers inspiration and materials to inform and support project work by students working individually or in groups. Teachers and tutors may lead by choosing materials which will extend and illuminate the curriculum tailored both to the specific requirements of the syllabus and to the level and interest of their students.

A Note on the Scope

The scope of the chemistry presented in this book has been tabulated for ease of reference by the reader, and is outlined in Table 1.1.

A Note on Nomenclature

Many of the natural products are given trivial chemical names, which are used throughout the book. However, where appropriate, we use the International Union of Pure and Applied Chemistry system to describe chemical formulae.

A Note on Botanical Names

Every plant belongs to a botanical family, and is further classified into its genus and species. The taxonomy of the plant is given a Latin name expressing both the genus and species, and is italicized.

A Note on References

We have included some selected reading material (mostly text books) for those wishing to delve further into the background of the materials we present. We also cite some primary references (from journals, usually found in major university libraries) which support key points.

TABLE 1.1
An Outline of the Chemistry Arising from the Natural Products Described in Each Section

Chapter	Section	Natural Products Chemistry	Organic Chemistry
2	**Medical Marvels**		
	Humble potato	Steroids	Alkanes, cycloalkanes
		Hormones	Benzene and aromaticity
			Industrial fractional distillation
	Willow bark	Substituted aromatic compounds	Carboxylic acids
		Salicylic acid and aspirin	Phenol and phenols
			Acetylation reaction
	Cinchona and artemisia	Quinine	Carbon–oxygen bonds
		Artemisinin	Oxygen–oxygen bonds
			Peroxides
	Foxglove	Glycoside lactones (cyclic esters)	Alcohols, acids and esters
	Periwinkle	Alkaloids	Functional groups
		Introducing indole alkaloids	Fractional distillation
		Vincristine and vinblastine	Acid–base extraction
	Pacific yew	Terpenes	Isomers
			Elimination reactions
			Stereochemistry
			Chirality
			NMR spectroscopy
3	**Modern Miracles of Foods and Ancient Grains**		
	Chia and quinoa	Lipids	Alkanes, alkenes
		Gluten sensitivity	Saturated and unsaturated fatty acids, transesterification
			Human health issues
			Vegetable oil and biodiesel
	Wheat	Lignan and lignin	Amino acids
			Peptide link
		Phenylpropanoids	Building blocks
		Fermentation	Condensation reaction and hydrolysis
		Shikimic acid pathway	
		Germination inhibition	Proteins and enzymes
			Crop yield, green issues
	Rice	Oxygen in the organic ring	Aldehydes (alkanals)
		Mono-, di- and polysaccharides	Ketones (alkanones)
		Anthraquinones, glycosides	Chromatography—paper, thin layer, gas and ion exchange
	Cordyceps	Role of NAD+ and NADH in redox reactions	Fermentation
			Sugars to alcohol and CO_2
			Redox reactions

(Continued)

TABLE 1.1 (*Continued*)
An Outline of the Chemistry Arising from the Natural Products Described in Each Section

Chapter	Section	Natural Products Chemistry	Organic Chemistry
	Garlic	Organo-sulfur compounds	Fossil fuels, air pollution
			Group VI elements of periodic table
			thiols
4	**Beverages**		
	Tea	Catechins	Phenolic ring
		Polyphenols (flavonoids)	Electrophiles
		Building blocks	Nucleophiles
			Substitution reactions
			Free radicals
			Antioxidants
	Cocoa	Flavonoids (polyphenols)	Redox reactions
			Antioxidation
			Free radicals
			Human health
	Coffee	Caffeine	Isolation of caffeine
		Cyclic aromatic amines	Extraction and purification
			Cyclic aromatic amines
			Resonant structures
			Zwitterions
	Maca	Alkaloids	Secondary metabolites
		Indole	Nitrogen in carbon rings
		Indole alkaloids	Acid–base extraction
5	**Euphorics**		
	Asian poppy	Morphine, codeine	Esters, alcohols and acids
		Heroin	Acetylation
	Cannabis	Cannabinoids	Isomers
		Terpenes	Polymerization from
		Isoprene rule	Elimination reactions
	Coca	Cocaine	Cyclic amines
		Alkaloids	Salt formation
	Tobacco	Nicotine—an alkaloid	Amines, amides
		Pyridine ring	Heterocyclic aromatics
		Pyrrole ring	Amines as building blocks
6	**Exotic Potions, Lotions and Oils**		
	Camphor	Terpenes—linear and cyclic	Polymerization
			Nucleophilic addition
			Carbonyl functional group
			IR spectroscopy
			Greenhouse gases
			Global warming
	Frankincense	Cyclic terpenoids	Steric hindrance
	and myrrh	Nomenclature, cyclic terpenes	Essential oils

(*Continued*)

TABLE 1.1 (*Continued*)
An Outline of the Chemistry Arising from the Natural Products Described in Each Section

Chapter	Section	Natural Products Chemistry	Organic Chemistry
	Lavender	Hydrosols and colloids	Fractional distillation
		Vegetable oils and fats	Steam distillation
		Glycerol	Soaps, esters, detergents reactions
		Fatty acids	
	Aloe	Cosmetic oils	Chromatography—paper, thin layer, gas, liquid, ion exchange
		Polysaccharides	
		Glycosides, anthraquinones	
		Galactomannan	Hydraulic "fracking"
7	**Colorful Chemistry: A Natural Palette of Plant Dyes and Pigments**		
	Green plants	Plant Kingdom	Photosynthesis
		Chlorophyll	Chromophores
		Pigments and dyes	Chelates and ligands
		Symbiosis	UV absorption spectroscopy
			Climate change
			Carbon accounting
	Saffron and yellow dyes	Carotenoids	Human health
		Carotenes	Vitamin A
		Xanthophylls	Carotenoid dyes
		Allyl functional group	Food color and flavoring
	Woad and indigo	Alkaloid dyes	Textile dyes
		Indigo, a purple-blue dye	Modern synthetic dyes
		Indigo white	Azo dyes
	Red dyes, henna, etc.	Quinones	Chromophores
		Naphthoquinones	Color fastness in dyeing
		Anthraquinones	Mordants
		Quinones as building blocks	Applications as dyes
	Reversible colors	Flowers, berries and fruit	Reversible dyes
		Flavonoid dyes	Color in flowers and leaves
		Anthocyanins	Acid/base indicators
		Phytochrome	Circadian and seasonal
			Rhythms, photoperiodism

SUGGESTED FURTHER READING

M. Balick, P. Cox. 1996. *Plants People and Culture: The Science of Ethnobotany*. Scientific American Library.

H. Hobhouse. 2005. *Seeds of Change. Six Plants That Transformed Mankind*. Counterpoint.

C. H. Howell. 2011. *Flora Mirabilis: How Plants Have Shaped World Knowledge, Health, Wealth and Beauty*. National Geographic.

P. Laszlo. 2007. *Citrus. A History*. University of Chicago Press.

J. L. Swerdlow. 2000. *Nature's Medicines. Plants That Heal*. National Geographic.

2 Medical Marvels

INTRODUCTION

Medicinal plants and plant-derived medicines are widely used in traditional cultures all over the world. People who use traditional remedies may not understand the scientific rationale for why they work, but know from personal experience that some plants can be highly effective. These plants were probably selected through trial and error.

Traditional medicine often aims to restore balance to the body. In contrast, Western allopathic medicine often uses a well-defined, single chemical entity with specific medicinal properties.

Nevertheless, many of our so-called "modern day" drugs have origins in ancient medicine. Furthermore, there has been a renaissance of awareness in the use of natural "alternative" medicines. The vast array of medicinal plants available from all parts of the world has stimulated much scientific and clinical interest which, in some instances, has provided significant commercial returns.

Each illustration included in Chapter 2 reveals how ancient civilizations survived, thrived and flourished by exploiting local plants. Subsequently, through a combination of trade, conquest, vision and ingenuity, the West successfully absorbed indigenous knowledge into its culture and economy. In more recent times, chemical examination of these plant extracts has led to the identification of many natural product constituents, which provide modern medications of inestimable value to humankind.

Ancient Medicine

Anecdotal and traditional wisdom concerning the use of botanical compounds is documented in the rich histories of traditional Chinese medicine (TCM) and of Ayurvedic medicine from India (and widely practised today in India), and in treatments used by Native Americans. In fact, medicinal plants and plant-derived medicines are widely used in traditional cultures all over the world.

HISTORICAL NOTE: AYURVEDIC MEDICINE

In Sanskrit, ayur means life or living, and veda means knowledge, so Ayurveda may be construed to mean "knowledge of living."

Ayurveda originated in the early civilizations of India some 3,000–5,000 years ago and is mentioned in the Vedas, the ancient religious and philosophical texts which are among the oldest literature surviving in the world.

Ayurveda involves living in harmony with nature and maintaining the human body so that mental and spiritual awareness may be promoted.

The goals of Ayurvedic medicine are prevention as well as promotion of the body's own capacity for maintenance and balance. Ayurvedic physicians claim that their methods help many stress-related, metabolic and chronic conditions. Ayurvedic medicine utilizes diet, purification techniques, herbal and mineral remedies, yoga, breathing exercises, meditation and massage therapy as holistic healing methods.

Historians believe that Ayurvedic ideas were transported from ancient India to China, and were instrumental in the development of Chinese medicine. Ayurvedic medicine is widely practised in modern India today and has been steadily gaining followers in the West.

Records of Ayurvedic medicine are a vast repository of knowledge in the application of herbal extracts to specific health problems, although much of this knowledge is yet to be fully explored and researched by modern scientific methods.

CENTRAL AMERICA'S HUMBLE POTATO!

Abstract: How a humble potato led to the genesis of the birth control pill. Few plants have had a greater impact on modern society. Indigenous to Central America and originally used as a staple food, this humble potato-like tuber contains the natural ingredients which transformed the world through the development of the modern birth control pill bringing about the profound social, cultural and economic impacts of oral contraception.

Natural Products Chemistry

- Steroids
- Hormones, estrogen and progestogen.

Curriculum content

- Hydrocarbons
- Alkanes and cycloalkanes
- Benzene and aromatic compounds
- The profound social, cultural and economic impacts of oral contraception.

Steroids

A steroid contains a characteristic arrangement of four cycloalkane rings, which are joined to each other. The main way in which steroids vary from one another is through the functional groups attached to the four-ring core (Figure 2.1).

Hundreds of distinct steroids are found in plants, animals and fungi. Examples of steroids include the dietary fat, cholesterol, and the sex hormone, testosterone.

Ancient Forms of Contraception

The desire to control contraception goes back to the dawn of culture. Women have risked their health and lives in child birth.

In Egypt, the Kahyn papyrus (1,850 BC) described the use of pessaries. They were made by mixing crocodile dung and selected herbs, which may have led to irritation and infection. However, the presence of a foreign object in the uterus may have

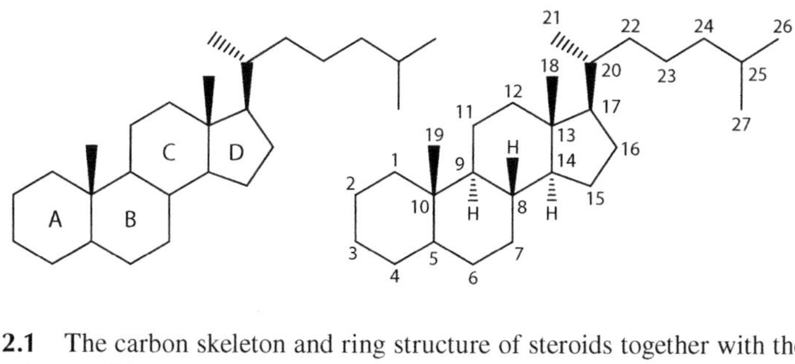

FIGURE 2.1 The carbon skeleton and ring structure of steroids together with the numbering system for steroids.

deterred any union between sperm and egg. (This idea forms the principle in modern times of the contraceptive use of a thin copper coil—the intrauterine device (IUD)). A waxy seal, forming around the cervix, may also have presented a physical barrier to sperm. However, it is unlikely that these early pessaries were entirely effective. Modern pessaries, known as vaginal suppositories, also contain wax-like material, which is infused with a spermicide. In ancient times, men used snake skins as condoms, whereas today, these physical barriers are made of latex or polyurethane.

It is also well documented that for centuries, women have used as an oral contraceptive an extract from the seeds of *Daucus carota* (known also as Queen Anne's Lace or Wild Carrot). The earliest references appear in a work written by the Greek physician, Hippocrates, in the fifth century BC. Modern studies performed on mice reveal that these extracts from the seeds block the production of the important hormone, progesterone (Figure 2.2). Progesterone is essential for pregnancy to occur, as its function is to prepare the uterine endometrium to receive an egg. If the egg does not implant then it will begin to break down and will no longer be viable. If the fertilized egg has been implanted for only a short period, it is believed by some that an extract from the Wild Carrot will cause it to be released. Plants such as Queen Anne's Lace (Wild Carrot) are still used in parts of rural United States as a "morning after" pill.

Another plant extract used as an oral contraceptive is from *Silphium*, a species of Ferula. *Silphium* was an essential item of trade in the ancient North African city of Cyrene where it was so important to the economy that many Cyrenian coins bore a picture of the plant.

Extracts from the plant Pennyroyal (*Hedeoma pulegoides* or *Mentha pulegium*) are also well known for their abortive effects. The extracts cause the uterine muscles to contract promoting menstrual discharge. Finally, the *Asafetida* species of plant has also been reported to have contraceptive and abortive activity. However, none of these plants became significant enough as oral contraceptives to have directly influenced the development of the modern birth control pill.

THE SEARCH FOR THE MODERN BIRTH CONTROL PILL: THE HORMONES, ESTROGEN AND PROGESTERONE

The modern combined oral contraceptive pill, often referred to as the birth control pill, or colloquially as "the pill," is a birth control method that includes a combination of an estrogen and a progestogen. When taken orally every day, these pills

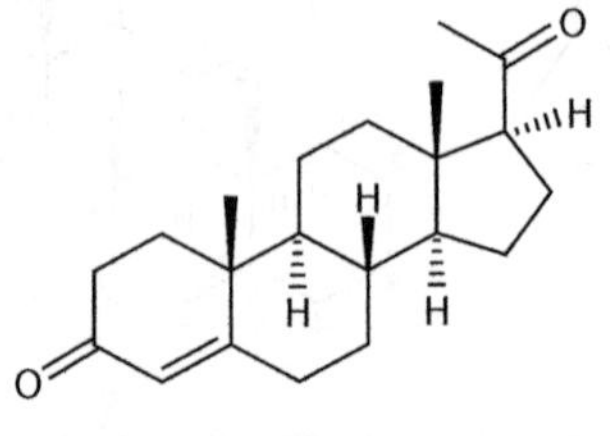

FIGURE 2.2 Chemical structure of progesterone.

inhibit female fertility. They were first approved for contraceptive use in the United States in 1960. They are a very popular form of birth control used by more than 100 million women worldwide, and by almost 12 million women in the United States.

The genesis of the birth control pill derives from the Mexican yam (*Dioscarea villosa* or *Dioscarea barbasco*). This potato is one of the first commercial sources of the key chemicals, known as steroid saponins. The yam is believed to originate from Central America, and approximately 800 *Dioscarea* species are currently known. The wild yam itself has no contraceptive value. Essentially, it provides the starting material from which the steroidal hormones are manufactured by modern technology through a combination of synthetic methods and also microbial transformation. However, in the 1930s, before the discovery of the usefulness of this yam, two important discoveries were made.

Firstly, scientists isolated and determined the structure of small amounts of steroid hormones. They then found that high doses of these hormones, in the form of androgens, estrogens or progesterone, inhibited mammalian ovulation. The isolation of these natural hormones was achieved by research on animal sources and carried out by the major European pharmaceutical companies. Since only small amounts were recovered they were extraordinarily expensive. However, it became immediately apparent that an urgent need for an abundant, reliable and cheap supply of steroids was needed.

Isolation of Natural Diosgenin from the Mexican Yam

In 1939, Professor Russell Marker, at Pennsylvania State University, developed a method of synthesizing progesterone from plant steroids, known as the sapogenins. Initially, he used sarsapogenin from sarsaparilla. However, his chemical route and methodology proved to be far too expensive on a commercial scale.

This finding led to investigation of a variety of Mexican yams, including *Dioscorea mexicana* and *Dioscorea barbasco*, which are found in the rain forests of Veracruz. The tuber (root) of the yam contains the natural chemical called diosgenin (Figure 2.3). This abundant compound could be used as the starting material in the synthesis of hormones on an industrial scale. Diosgenin can be easily converted chemically by opening the rings containing C21 to C27 as shown in Figure 2.1. This is then followed by further degradation of the molecule to reach the structure shown in Figure 2.2.

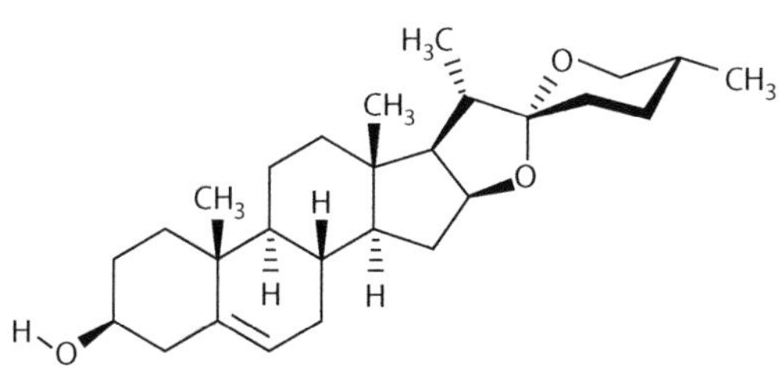

FIGURE 2.3 The chemical structure of diosgenin.

Although by midway through the twentieth century, the stage appeared set for the development of an oral hormonal contraceptive, pharmaceutical companies, universities and governments showed little interest in pursuing further research. At this time in 1944, having developed a synthesis of progesterone from diosgenin from the Mexican yam, Professor Marker left Pennsylvania State University to found a new company, Syntex, in Mexico City. At Syntex, Marker continued to perfect the extraction of the saponins from *Dioscorea mexicana* and then the manufacture of the key hormones. Importantly, due to this achievement, the monopoly of the European pharmaceutical companies was broken, which had until that time controlled production of steroid hormones. As a consequence, the price of progesterone fell dramatically by almost 200-fold over the next 8 years.

Chemical Magic in the Laboratory: Synthesis of Norethindrone

In 1951, the combined brilliance of three extraordinary chemists at Syntex, Carl Djerassi, Luis Miramontes and George Rosenkranz, led to the synthesis of the first oral, highly active progestin, namely, norethindrone (Figure 2.4). It should be noted that this synthetic hormone is a variation of natural progesterone. Furthermore, in

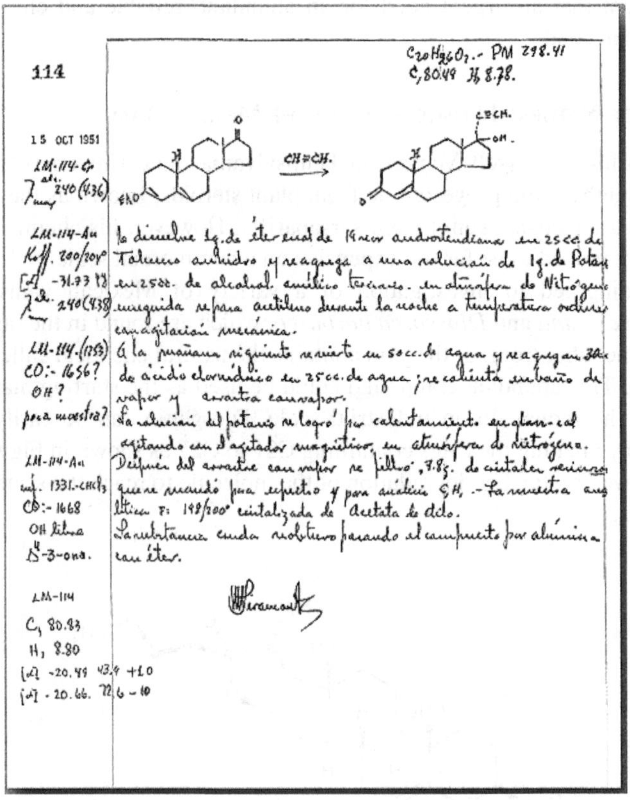

FIGURE 2.4 A copy of the original page from Luis E. Miramontes' laboratory notebook, signed October 15, 1951. (Courtesy of the late Carl Djerassi.)

the following year, at the competing Searle Company in Skokie, Illinois, their chemist, Dr. Frank Colton synthesized the orally active progestin known as norethynodrel (an isomer of norethindrone) followed in 1953 by norethandrolone.

Biological Studies of Progesterone to Prevent Ovulation

Stimulated by a steady supply of research-grade chemical material, important biological studies on the activity of progesterone in inhibiting ovulation progressed rapidly. In early 1951, the reproductive physiologist, Dr Gregory Pincus, who was at the time a leader in hormone research, showed that injection of progesterone suppressed ovulation in rabbits.

Also, significantly in this pivotal year, Pincus met with Margaret Sanger, the founder of the American birth control movement (see Historical Note).

HISTORICAL NOTE: THE IMPACT OF MARGARET SANGER

The importance of Margaret Sanger, a nurse and a huge advocate of female contraception, cannot be overlooked. Her crusade to legalize birth control spurred the movement for women's liberation.

Margaret Sanger became acutely aware of the effects of unplanned and unwelcome pregnancy during her work with poor women on the Lower East Side in New York. Sanger had witnessed how her own mother's health had suffered as she bore 11 children. Sanger realized the importance to women's lives and health of the availability of birth control. In 1912, Sanger gave up her nursing work to dedicate herself full time to the distribution of birth control information. However, due to the passage of the Comstock Act of 1873, it was still forbidden to distribute either birth control devices or information. Sanger continued to write articles on health for the Socialist Party paper, "The Call," and published articles: "What Every Girl Should Know" (1916) and "What Every Mother Should Know" (1917). However, she was indicted for "mailing obscenities" and fled to Europe. Eventually, the indictment was withdrawn.

During World War I, Sanger set up the first birth control clinic in the United States yet she was sent to the workhouse for "creating a public nuisance" and was arrested and prosecuted many times. The resulting public outcry helped lead to changes in the law which in turn empowered doctors to give birth control advice to their patients.

In 1927, Sanger helped to organize the first World Population Conference in Geneva.

In 1942, after several organizational mergers and name changes, The Planned Parenthood Federation came into being.

Medical Approval and Social Acceptance

By 1954, studies had advanced on the ovulation-suppressant potential of progestins, which were administered orally. The first medical trials of an oral contraceptive, later commercially known as Enovid, began in 1956 in Puerto Rico, which led the American Food and Drug Administration (FDA) to approve Enovid for menstrual disorders; and by May 1960, the FDA gave approval for its use as a contraceptive.

Use of the birth control pill varies widely, country by country and by age, education and marital status. One-third of women aged 16–49 in Great Britain use either the combined pill or a progestogen-only version, whereas only 1% of women use the pill in Japan. In Japan, lobbying from the Japan Medical Association prevented the birth control pill from being approved for nearly 40 years. There were concerns: safety of the drug over the long term, and that the use of the pill could lead to diminished use of condoms and thereby to potential for a rise in the rate of sexually transmitted infections. As of 2004, condoms accounted for 80% of birth control use in Japan, which may explain the comparatively low rate of AIDS in the country. By 1999, the pill was finally approved for use in Japan. However, according to estimates, only 1.3% of Japanese females use the pill compared with 15.6% in the United States.

The Profound Social, Cultural and Economic Impacts of Reliable, Oral Contraception

Since the introduction of the oral birth control pill in 1960 and its approval by the FDA, its use has spread rapidly, generating enormous social impact. Since the pill allowed women to have a sexual relationship while pursuing a career, it became a significant factor in a quiet revolution. Many consider it to be the most socially significant medical advance of the twentieth century. The birth control pill has helped women gain more control of their lives and has altered the nature of the nuclear family and life profoundly. Many economists argue that the availability of the birth control pill led directly to an increase in the proportion of women in the labor force and was a key influence in determining the modern economic role of women. It is notable that after the birth control pill was legalized there was a sharp increase in college attendance and in graduation by women. Family planning allowed women to make long-term educational and career plans. The pill offered opportunity to delay the timing of marriage allowing women to invest in education and in other forms of human capital and to become more career oriented.

Owing to the fact that the birth control pill was inexpensive and effective, widespread adoption changed the nature of debate over premarital sex and promiscuity. Never before had sexual activity been so divorced from human reproduction. In this regard, the proliferation of the use of oral contraceptives has required religious authorities to reexamine the relationship between sexuality and procreation.

Transformation: Global Emergence of the Yam

In 1960, two million women were using the pill and over 100,000 Mexican peasants were gathering the raw material used in its production. In order to meet the demand, more than 10 tons of wild yam were removed each week at extraordinarily low prices from the areas around Oaxaca, Veracruz, Tabasco and Chiapas in Mexico. Scientists relied on local indigenous knowledge to cultivate and harvest the plant. Yams made their way from the Mexican jungles to domestic and foreign laboratories and into the medicine cabinets of millions of women around the world. At the time, little recognition was afforded to Mexican peasants who labored for almost 30 years to collect the yams, yet had no sense of its value in the marketplace or of the importance of their contribution.

Interest in the Mexican yam could no longer be confined within national borders as growing pressure arose from a combination of continued progress in chemistry research, improved pharmaceutical technology and changes in the social and political outlook across the world. The Mexican government eventually established a state-owned company in 1975 to compete with foreign laboratories. Funds were thus secured for the training of scientists and the development of a stronger domestic pharmaceutical industry in Mexico.

Arguably, the indigenous, poor, uneducated yam pickers represented in many respects the antithesis of modernity, but they became an essential link in finally introducing to Mexico a modern, domestic industry-patented medications. In this particular case, an alliance of science and farming practice resulted in a reshuffling of social hierarchy in rural Mexico, and gave real monetary value to an otherwise low-value crop.

HYDROCARBONS

Table 2.1 illustrates the general classification of hydrocarbons and provides examples.

SATURATED HYDROCARBONS

Alkanes

Alkanes are described as saturated hydrocarbon molecules because each carbon atom, which has a valency of four, is bound covalently to four other atoms—either carbon or hydrogen.

Alkane molecules can be in straight chains of carbon atoms (see examples in Table 2.2) or in branched chains.

Cycloalkanes

Cycloalkanes exist too, although they are usually in the form of six-membered carbon rings or larger. Owing to the ring structure, however, some strain exists in the

TABLE 2.1
The General Classification of Hydrocarbons

Hydrocarbons						
Chain			Cyclic			
Saturated	Unsaturated		Carbocyclic		Heterocyclic	
Alkanes	**Alkenes**	**Alkynes**	**Alicyclic**	**Aromatic**	**Alicyclic**	**Aromatic**
Methane	Ethene	Ethyne	Cyclohexane	Benzene	Cyclohexylamine	Pyridine

TABLE 2.2
Examples of Hydrocarbons with the General Formula C_nH_{2n+2}, Known as Alkanes

Formula	Name	Physical Properties	Boiling Point	Melting Point
CH_4	Methane	Colorless gas		
C_2H_6	Ethane	Colorless gas		
C_3H_8	Propane	Colorless gas		
C_4H_{10}	n-Butane	Colorless gas		
C_5H_{12}	n-Pentane	Colorless, volatile liquid	36°C	
$C_{18}H_{38}$	n-Octadecane	White solid		28°C

FIGURE 2.5 The structure of a cyclohexane molecule.

bonds. For instance, flat planar molecules of cyclopropane can be formed, but they are very unstable, whereas in cyclohexane (Figure 2.5) the carbon–carbon bonds are formed at the usual tetrahedral angle of 109.5 degrees. As a consequence, the cyclohexane ring is puckered into what is commonly referred to as either the chair or the boat form.

Owing to the ring structure, cycloalkanes have two fewer hydrogen atoms than the corresponding alkane, and have the general formula C_nH_{2n}. However, the physical and chemical properties of cycloalkanes are scarcely different to those of the corresponding straight chain or branched chain alkane.

Alkanes and cycloalkanes, in gaseous or vapor form, are readily oxidized in the presence of oxygen or air with the release of a great deal of heat energy per mole.

$$C_3H_8 + 5O_2 = 3CO_2 + 4H_2O$$

A mole is defined as the molecular weight of a substance expressed in grams based on the standard of 12 g of carbon 12.

In the presence of a halogen and energized by light, alkanes undergo substitution reactions:

$$CH_4 + Br_2 = CH_3Br + HBr \text{ then } CH_3Br + Br_2 = CH_2Br_2 + HBr$$

and so on to CBr_4.

Crude Oil and Industrial Fractional Distillation

Crude oil or petroleum is a highly complex mixture of alkanes and many other organic compounds. Petroleum is found in underground reservoirs in rock strata

TABLE 2.3
Examples of Oils That Can be Refined by Fractional Distillation

Fraction	Boiling Range	Carbon Atoms per Molecule	Use
Light petroleum	20–90°C	4–6	Solvent
Gasoline (petrol)	100–200°C	8–12	Motor fuel
Paraffin, kerosene	200–300°C	12–16	Diesel fuel
Oil	Above 300°C	More than 25	Lubrication
Bitumen	Solid residue	Large numbers	Road construction

where it was formed by the very slow decomposition of organic matter from plants and animals in the absence of air under the influence of great heat and pressure. The composition of crude oil varies considerably from place to place; some deposits being dominated by straight and branched chain alkanes, whereas other sources contain greater proportions of cyclic alkanes, which often have carbon chain branches.

Crude oil is fractionally distilled and different fractions are collected on an industrial scale (Table 2.3) in oil refineries (Figure 2.6).

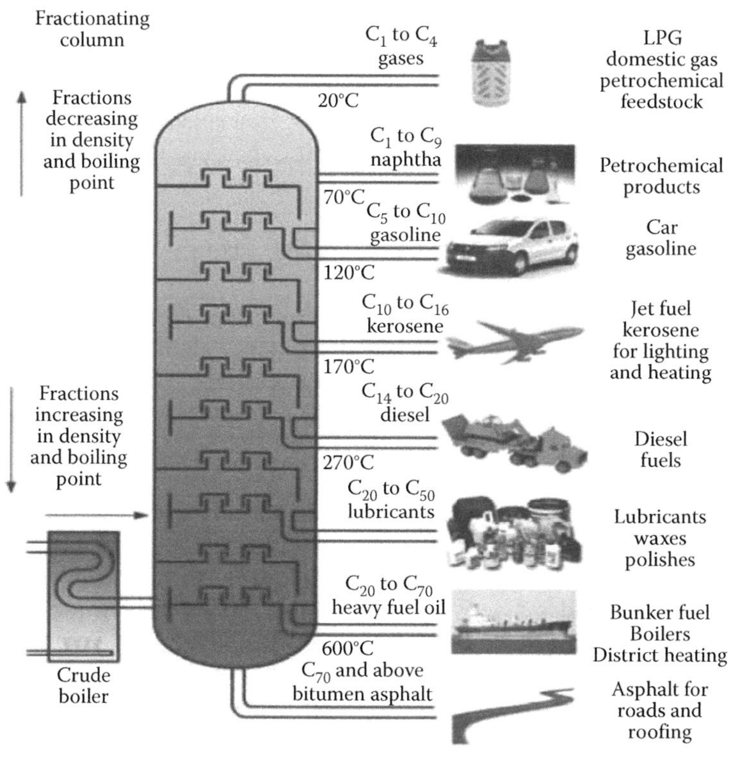

FIGURE 2.6 **(See color insert.)** Diagram of an industrial fractionating column. (With permission under terms of GNU Free Documentation License.)

Benzene and Aromatic Compounds: Physical and Chemical Properties

Benzene is a colorless, volatile liquid, and is a recognized carcinogen.

The molecular formula, C_6H_6, reveals that there is a high percentage of carbon, and that benzene is, to some degree, unsaturated. Indeed, benzene is described as the simplest of the aromatic hydrocarbons.

Studies of the structure of a molecule of benzene have shown that the molecule is in the form of a planar, hexagonal ring of six carbon atoms with the internal angle between them being 120 degrees. This is in marked contrast to the molecular structure of cyclohexane which is discussed earlier in the chapter. The chemical reactivity of benzene is also revealing in that it readily undergoes addition reactions and behaves as a nucleophile in substitution reactions (see also the section on "Tea, from Legend to Healthy Obsession" in Chapter 4 and "A Plant from the East Indies, Camphor" in Chapter 6).

At ordinary temperatures and elevated pressure, benzene can be hydrogenated in the presence of a finely divided catalyst of platinum to produce cyclohexane.

$$C_6H_6 + 3H_2 = C_6H_{12}$$

Another example of an addition reaction involves halogens such as chlorine, which in the presence of the energy source, ultraviolet light, will yield benzene hexachloride.

The replacement of a hydrogen atom in benzene occurs much more readily than the corresponding replacement in an alkane or cycloalkane. In the presence of concentrated nitric and sulfuric acids at about 50°C, benzene will react to give nitrobenzene by nucleophilic substitution.

$$C_6H_6 + HNO_3 = C_6H_5NO_2 + H_2O$$

Halogenation of benzene provides another example of nucleophilic substitution. The reaction takes place in the presence of a catalyst, aluminium chloride, when chlorine is passed through the liquid at room temperature.

$$C_6H_6 + Cl_2 = C_6H_5Cl + HCl$$

Chlorobenzene is known to be the result when a measured increase in weight has occurred.

The Friedel–Crafts reaction is an important example of a nucleophilic substitution which provides a very useful synthetic pathway in organic chemistry. Catalyzed by aluminium chloride, benzene can undergo alkylation or acylation to produce an alkyl benzene (such as toluene) or an acyl benzene (such as acetophenone).

$$C_6H_6 + CH_3Cl = C_6H_5 \cdot CH_3 + HCl$$

$$C_6H_6 + CH_3 \cdot CO \cdot Cl = C_6H_5 \cdot CO \cdot CH_3 + HCl$$

Reference to the sections on "Europe Solves a Headache" in Chapter 2, and "Morphine: A Two-Edged Sword" in Chapter 5, is also advised for more reading on alkylation.

Finally, mention must also be made of the violent oxidation reaction between benzene vapor and ozone, or indeed between any hydrocarbon and ozone, which ultimately yields carbon dioxide and water. This property of ozone has crucial consequences in the upper atmosphere of the Earth where hydrocarbon pollutants, arising from aircraft at high altitude or from upward diffusion of propellant gases from aerosol cans or refrigeration systems in use in the lower atmosphere, have caused serious reduction in the partial pressure of ozone—commonly referred to as the ozone hole. A key property of ozone is that it absorbs high energy, short wavelength ultraviolet radiation. If ultraviolet radiation is not reduced in intensity to natural levels at the surface of the Earth, it can cause serious damage to plant and animal life.

THEORY OF THE MOLECULAR STRUCTURE OF BENZENE

The structure of benzene with a molecular formula of C_6H_6 has always presented something of a problem. If the molecule were linear, that would suggest that benzene ought to have a similar degree of unsaturation to that of ethyne or acetylene, but benzene is much less reactive in addition and substitution reactions. It is more stable than anyone would expect at first sight. In 1865, Kekule was the first person to suggest a cyclic structure which might be based on a hybrid of resonant forms (Figure 2.7). This idea has been reinforced somewhat by theory in quantum mechanics.

Modern spectroscopic studies of the benzene molecule indicate unequivocally a planar, regularly hexagonal molecule with C–C–C and C–C–H bond angles of 120 degrees. These facts, together with the relative stability of benzene, compared to alkenes and alkynes, have led to the acceptance of a molecular orbital theory in which the p electron orbitals of neighboring carbon atoms overlap one another, thus allowing electrons to pass around the ring above and below the plane of the ring. These are the delocalized electrons which are considered responsible for the aromatic character of benzene and larger compounds containing benzene as a building block. As noted in the section on "Tobacco: A Profound Influence on

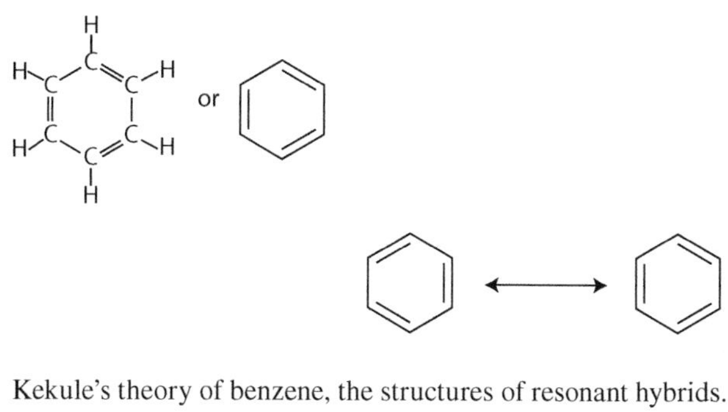

FIGURE 2.7 Kekule's theory of benzene, the structures of resonant hybrids.

FIGURE 2.8 The delocalized electron structure representing a molecule of benzene.

the World" in Chapter 5, some heterocyclic compounds, such as pyridine, pyrrole and thiophen, are also aromatic for the same reason—the delocalization of electrons around a ring molecule, above and below its plane (Figure 2.8).

While Kekule's structures are not quite consistent with physical and chemical evidence, they are still used today in depicting possible mechanisms of reactions of benzene derivatives, especially when hydrogen atoms in the ring are substituted in the ortho- and para-positions. An example of this behavior is found in the molecule of phenol, which behaves as a weak acid in aqueous solution and reacts much more easily than benzene does in electrophilic substitution reactions involving the 2 and 4 positions (ortho and para) in the carbon ring (see also the section on "Tea, from Legend to Healthy Obsession" in Chapter 4 for more on phenol).

Summary of the Characteristics of Aromatic Compounds

a. Aromatic compounds burn in air with a luminous, smoky flame, due to the high proportion of carbon in the molecule.
b. These compounds are somewhat less reactive toward oxidation and reduction than alkenes and alkynes.
c. They do not undergo addition reactions very readily.
d. Aromatic compounds behave as weak nucleophiles, and are readily involved in substitution reactions with electrophiles.

QUESTIONS

1. Give an account of the process of refining petroleum making sure to explain the science involved in each step including fractional distillation and cracking.
2. All manufactured goods have a carbon footprint. Choose an example of a product and explain how the carbon footprint of the product arises from start to end, from the conversion of raw materials to the finished product appearing in the shops.
3. Taking as an example the steroid, diosgenin, describe the range of chemical reactions which, in principle, you would expect to arise from the functional groups present; an ether-like linkage, a carbonyl group, a cyclohexane ring and a degree of unsaturation in the molecule. Describe any limitations in practice.
4. Give a full account of the implications of the high reactivity of introduced hydrocarbons with naturally occurring ozone in the stratosphere of the Earth's atmosphere. Explain what international measures have been

undertaken to reduce the release of hydrocarbons to counteract depletion of the ozone layer.
5. Ozone can appear as a pollutant at low level in the Earth's atmosphere (or troposphere) in the presence of strong sunlight resulting in a photochemical smog which is injurious to human health. Explain carefully all of the factors involved, and especially the chemistry, which give rise to this phenomenon in developed countries of the world.
6. Compare and contrast the physical and chemical properties of cyclohexane and benzene.
7. Draw together the evidence for the delocalized ring structure of benzene molecules and relate this to the Kekule model of resonant hybrids, and also to modern molecular orbital theory.
8. The simplest polycyclic aromatic hydrocarbon is naphthalene, $C_{10}H_8$, which is a white crystalline solid. It is the main constituent of mothballs. Given the knowledge of benzene, draw a delocalized structure of a molecule of naphthalene and describe the range of chemical properties you would expect naphthalene to have.

REFERENCES

J. W. Goldzieher and H. W. Rudel. 1974. How the oral contraceptives came to be developed. *The Journal of the American Medical Association* 230(3):421–425.

G. Pincus. 1958. The hormonal control of ovulation and early development. *Postgraduate Medicine* 24(6):654–660.

SUGGESTED FURTHER READING

N. Applezwei. 1962. *Steroid Drugs vii–xi*. Pages 9–83. McGraw-Hill.

R. Cooper, G. Nicola. 2014. *Natural Products Chemistry: Sources, Separations and Structures*. CRC Press, Taylor & Francis Group.

C. Djerassi. 2001. *This Man's Pill Reflections on the 50th Birthday of the Pill*. Pages 11–62. Oxford University Press.

J. M. Riddle. 1992. *Contraception and Abortion from the Ancient World to the Renaissance*. Page 28 and references therein. Harvard University Press.

E. W. Straus, A. Strauss. 2006. *Medical Marvels: 100 Greatest Advances in Medicine*. Pages 139–143. Prometheus Books.

EUROPE SOLVES A HEADACHE! EMERGENCE OF ASPIRIN

Abstract: The use of willow bark to relieve symptoms of the ague by riverbank communities led to the eventual development of the great miracle drug, aspirin, by the German pharmaceutical giant, Bayer Company.

Natural products chemistry

- Salicylic acid and aspirin.

Curriculum content

- Carboxylic acids
- Phenol and the hydroxyl group in an aromatic ring
- Acetylation of the hydroxyl group of salicylic acid to form aspirin.

SALICIN AND SALICYCLIC ACID

The active ingredient in willow bark is salicin (Figure 2.9) which is converted naturally within the human body into salicylic acid. Salicin is a glycoside that is revealed when it is hydrolyzed to salicylic alcohol and glucose (see also the sections on "Global Aloe" and "A Steroid in Your Garden" in Chapter 6 for more about glycosides).

The molecular structure of salicylic acid is shown in Figure 2.10. This molecule is an example of a bifunctional organic compound, whereby the chemical reactivity may be due to either the carboxylic acid group or the phenolic group or both.

CARBOXYLIC ACIDS

Carboxylic acids contain the carboxyl functional group –COOH, and examples are shown in Table 2.4. Nomenclature, as usual, follows the name of the stem of the corresponding alkane or benzene with the suffix *oic* applied. There are, of course, straight chain carboxylic acids with branched chain isomers, molecules with two or more carboxylic functional groups and aromatic carboxylic acids.

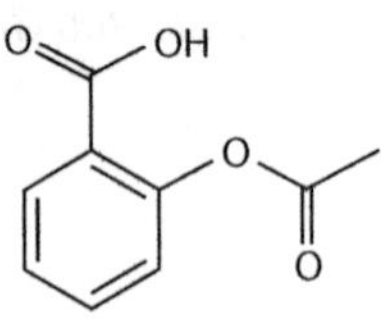

FIGURE 2.9 Chemical structure of salicin.

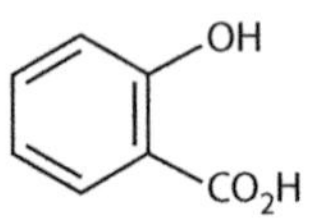

FIGURE 2.10 Molecular structure of salicylic acid.

TABLE 2.4
Examples of Simple Organic Carboxylic Acids

$HCOOH$	Methanoic acid (unsystematic name, formic acid)
$CH_3 \cdot COOH$	Ethanoic acid (unsystematic name, acetic acid)
$CH_3 \cdot CH_2 \cdot COOH$	Propionic acid
$CH_3 \cdot CH_2 \cdot CH_2 \cdot COOH$	Butanoic acid (unsystematic name, butyric acid)
$CH_2Cl \cdot CH(CH_3) \cdot COOH$	3-Chloro-2-methyl butanoic acid
$C_6H_4(COOH)_2$	Benzene dicarboxylic acid (informal name, phthalic acid)

Methanoic acid (b.p. 100°C) and ethanoic acid (b.p. 118°C) are both volatile liquids at ambient temperature and pressure and have pungent smells. The other acids are generally colorless crystalline solids.

Owing to the electronegativity of oxygen atoms relative to those of hydrogen, the hydroxyl bond is polar. This fact accounts for a number of the properties of carboxylic acids, especially those of low molecular mass where the hydrocarbon moiety in the molecule is less influential.

Due to the influence of hydrogen bonding in the liquid state or in aqueous solution, the lighter aliphatic acids are quite soluble in cold water, while aromatic acids are only sparingly soluble under the same conditions.

As the name of the class strongly suggests, these compounds

- Are acidic in aqueous solution
- Form organic salts readily in dilute solutions of inorganic bases
- Displace carbon dioxide from inorganic alkali metal carbonates
- Form ammonium salts or amides with ammonium hydroxide
- Combine with alcohols to form esters (catalyzed in aqueous solution by an inorganic acid).

More on the formation, properties and uses of esters is to be found in the section "A Steroid in Your Garden" in Chapter 2.

Phenol and Phenolic Compounds

Phenol (Figure 2.11) consists of a benzene ring in which one of the hydrogen atoms has been replaced by a hydroxyl group. More on the properties and uses of phenol and phenolic compounds is to be found in the section "Tea: From Legend to Healthy Obsession!" in Chapter 4.

Phenolic molecules dissolve sparingly in water as the polar hydroxyl group is able to form hydrogen bonds with water molecules. More importantly though, because of

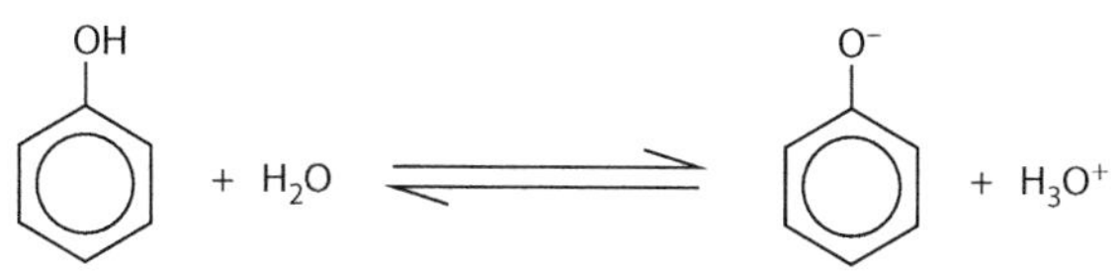

FIGURE 2.11 Partial dissociation of phenol in aqueous solution involving ions and the molecule in equilibrium.

stabilization of the anion through the delocalization of electrons over the benzene ring, the hydroxyl group in phenol partially dissociates in aqueous solution to form phenoxide anions and hydrogen cations.

As a consequence, the molecules and ions are in dynamic equilibrium and aqueous solutions of phenol have the properties of a weak acid:

- Reacting with strong bases to form a salt and water
- Unable to react with weaker bases such as sodium carbonate.

Even though phenol is different from aliphatic alcohols in many ways, phenol will form esters with a carboxylic acid. As carboxylic acids are weak acids themselves, the reaction is slow, but as an example, may be speeded up acceptably by using either an acyl chloride, such as ethanoyl chloride, or alternatively and more safely, ethanoic anhydride. Phenyl ethanoate would be the product.

$$CH_3 \cdot COCl + C_6H_5 \cdot OH = CH_3 \cdot COO \cdot C_6H_5 + HCl$$

Phenol is an important building block found in many natural products. Ease of electrophilic substitution (see Glossary) in the ortho- and/or para-positions (also known as the 2 and 4 carbon positions) of the aromatic ring is a notable chemical property of phenol. The directing effect of the hydroxyl group is so strong in phenol that in a chemical preparation it is often difficult to control the reaction to just monosubstitution.

The second main type of reaction which phenol undergoes involves replacement of the hydroxyl group by a carboxyl group, or by an ether linkage, or by a methyl or acetyl group.

More on the properties and uses of phenol and phenolic compounds is to be found in the section "Tea: From Legend to Healthy Obsession!" in Chapter 4.

Acetylation of the Hydroxyl Group of Salicylic Acid to Form Aspirin

Acetylsalicylic acid, commonly known as aspirin, may be prepared in the laboratory by substitution of the hydrogen atom of the phenol group of salicylic acid with an acetyl group (Figure 2.12). Further reference should also be made to the section titled "Morphine: A Two-Edged Sword" in Chapter 5 where the acetylation reaction is examined closely. Substitution of the phenol group may be achieved under anhydrous conditions using an acyl chloride, ethanoyl chloride, or preferably by using an acid anhydride, ethanoic anhydride, since the hazardous by-product, gaseous hydrogen chloride, is avoided.

Aspirin is a colorless crystalline solid (melting point 135°C), which is soluble in water. The drug is a well known as an analgesic (pain killer) and as a pyretic (reduces

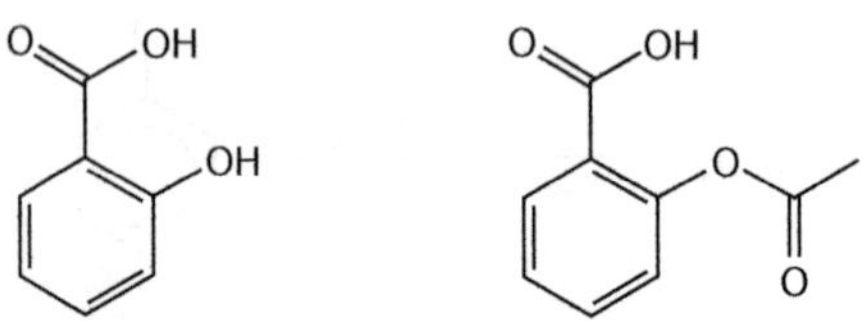

FIGURE 2.12 Salicylic acid and acetylsalicylic acid commonly known as aspirin.

fever or body temperature). Besides aspirin, another derivative of salicylic acid deserves mention. Methyl salicylate, or oil of wintergreen, occurs in many plants. Owing to its fragrant smell, methyl salicylate is used in perfumery.

HISTORICAL NOTE

From the earliest times, it was known that the chewing of the bark of the willow tree, *Salix alba*, reduced fever and inflammation. The ancient Greek physician, Hippocrates, is recorded as having recognized its therapeutic benefits.

In 1763, the Reverend Edmund Stone of Chipping Norton, Oxfordshire, England, read an obscure paper to the Philosophical Society of London entitled "An Account of the Success of the Bark of the Willow in the Cure of Agues." Later, he wrote about this lecture in a letter to the Right Honourable George, Earl of Macclesfield, who was the President of the Royal Society at the time.*

In the 1800's, pharmacists created salicylic acid in its acetylated form (acetylsalicylic acid)—more commonly known by its brand name, Aspirin. Aspirin was first isolated and synthesized by Felix Hoffmann, a chemist with the German company, Bayer, and marketed in 1897. The most widely used drug in the world continues to be Aspirin which remarkably has remained essentially unchanged for over 2,500 years.

* *Philosophical Transactions* 1683–1775. 53(1763):195–200. Published by the Royal Society.

QUESTIONS

1. Give the systematic names and molecular structures of the isomers of pentanoic acid.
2. Name the three isomers of benzene dicarboxylic acid, and give their structural formulae.
3. Give the systematic names of oxalic acid, $(COOH)_2$; lactic acid, $CH_3 \cdot CH(OH) \cdot COOH$; tartaric acid, $[CH(OH) \cdot COOH]_2$; benzoic acid, $C_6H_5 \cdot COOH$; and cinnamic acid, $C_6H_5 \cdot CH{:}CH \cdot COOH$.
4. Explain the origin of hydrogen bonding in carboxylic acids, and the nature of dimers in glacial ethanoic acid. Account, in turn, for the differences in the physical properties of low molecular weight carboxylic acids, high molecular weight carboxylic (fatty) acids and aromatic carboxylic acids.
5. Knowing the properties of a carboxylic acid and phenol, describe in full the chemistry of salicylic acid given that it is an example of a bifunctional organic compound.
6. Give examples of plants which are sources of oil of wintergreen, and explain how it is extracted and used commercially.
7. In phenol, the ortho- and para-positions in the carbon ring have been revealed in this section. Identify the *meta* position in the ring.
8. Explain why the OH group in phenol has such a strong directing effect within the carbon ring in reactions involving electrophiles.

REFERENCE

'Philosphical Transactions' 1683–1775, Vol. 53 (1763), pp. 195–200. *Published by the Royal Society.*

SUGGESTED FURTHER READING

D. Jeffreys. 2004. *Aspirin, Story of a Wonder Drug.* Bloomsbury.
C. C. Mann, M. L. Plummer. 1991. *Aspirin Wars.* A. A. Knopf.
E. W. Straus, A. Strauss. 2006. *Medical Marvels: 100 Greatest Advances in Medicine.* Prometheus Books.

ATTACKING MALARIA: A SOUTH AMERICAN TREASURE (BUT NOT GOLD) AND A CHINESE MIRACLE

Abstract: The huge cultural and economic impact of an extract from cinchona bark upon European colonization of tropical and subtropical regions of the world can scarcely be exaggerated.

Spanish colonists learned about cinchona from the native South Americans, brought the plant back to Europe and, as a consequence, reduced the spread of malaria. Later, scientists discovered how to isolate and refine the active biological ingredient, now known by the name, quinine.

Another botanical miracle emerged in recent times in China where initially secret research led to the development of a new antimalarial drug called artemisinin.

Natural products chemistry

- Quinine
- Artemisinin.

Curriculum content

- Carbon—oxygen bonds in organic compounds
- Oxygen—oxygen bonds in organic compounds
- Peroxides.

MALARIA

Malaria is a vector-borne disease transmitted by the bite of a female mosquito infected with single-celled (protozoan) parasites (known as plasmodium). The tiny parasites pass through the blood stream of the human victim and travel to the liver where they mature and reproduce before affecting the whole body by attacking red blood cells. Symptoms of malaria typically include headache, feverishness and fatigue. If not treated, malaria can cause death. The areas in the world most associated with malaria are shown in red in Figure 2.13.

Two major drugs are employed in the fight against malaria, and both originate directly from natural products; quinine from the bark of the cinchona tree found in South America; artemisinin from the leaves of *Artemisia annua* native to China. The latter discovery is particularly important, as the effectiveness of drugs based solely on quinine has gradually diminished as the infecting parasites have developed resistance to the quinine-based drugs. Subsequently, artemisinin has become the treatment of choice for malaria. However, the World Health Organization (WHO) called for cessation in the use of single doses of artemisinin in 2006 in favor of combinations of artemisinin together with another malaria drug in order to reduce the risk of the parasites developing resistance. Thus, artemisinin is usually combined with a synthetic derivative of quinine known as chloroquinine. In this dual dose, the drugs reinforce one another in addressing malaria in that they have complementary roles; the former is quick acting whilst the latter reduces inflammation. However, it remains to be seen whether the strategy of combination therapy will be entirely successful in the management of malaria.

FIGURE 2.13 **(See color insert.)** Parts of the world where malaria is endemic are shown in red on the map from the 2013 Global Malaria Mapper. (Courtesy of the World Health Organization (WHO), http://www.who.int/malaria/publications/world_malaria_report/global_malaria_mapper/en/)

Cinchona

Cinchona is an evergreen shrub or small tree indigenous to the high Andes of South America.

The botanical name of the genus Cinchona was given by Linnaeus in 1742 from the Indian name, *Quinaquina,* derived from the Quechua language of Ecuador, Peru and Bolivia. The Quechua people used an extract from the bark as a relaxant for muscles to combat extreme cold in the high mountains. It was also noticed that the extract had a favourable impact on malaria. Later use of an extract from the bark of the cinchona tree led to significant reduction in malaria among the ranks of conquistador invaders, and helped strengthen the might of Spain for generations. Because of its medicinal property, Spanish colonists brought the shrub back to Europe around the early 1600s. Later research expeditions left from France, The Netherlands and Great Britain. Eventually, the Dutch and British cultivated cinchona for this "mysterious extract" in the East Indies and in the Indian subcontinent, respectively.

Isolation of Quinine

This substance was first isolated in 1820 by Pierre Pelletier from the bark of the cinchona shrub. He separated a yellow gum with a bitter taste, which he called quinine. However, the exact chemical structure presented in Figure 2.14 was not fully elucidated until much later. Even today, quinine is still recognized as one of the most effective drugs in the treatment of malaria although details of the mechanism of its disruption of the life cycle of the plasmodium parasite are not understood.

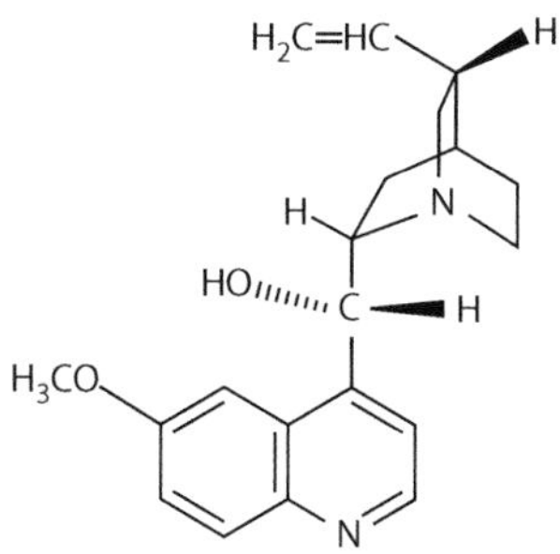

FIGURE 2.14 The structure of quinine, the antimalarial drug from the cinchona bark.

The quinine molecule has two distinctive parts; an aromatic component and an amine component. As a consequence, quinine

- Is soluble in water at room temperature to give an alkaline solution (pH 8.8)
- Readily forms salts with strong acids, examples being quinine sulphate and quinine hydrochloride.

Synthesis of Quinine in the Laboratory

The famous and brilliant British chemist, William Perkin, tried to synthesize quinine in 1856 but was unsuccessful. Eventually, quinine was synthesized in 1944. Quinine was the first drug from a natural product to be synthesized although the route was so complex it could not be made into a commercial success. Consequently, quinine is still obtained directly from the cinchona bark. Nowadays, quinine is chemically modified synthetically to make quinine derivatives, which are even more powerful as antimalarial drugs.

However, it should be remembered that Perkin's persistence in the early research of 1856 led remarkably to the discovery and production of a mauve-colored compound, which he established as a derivative of aniline. Inadvertently, he had fashioned the world's first synthetic dye, which became the basis of the entire aniline dye industry. For more on dyes, see the section on "Woad (Isatis tinctoria) and Indigo" in Chapter 7.

Uses of Quinine

Beyond its application as an antimalarial drug, quinine is widely utilized in the modern drinks industry. By repute, British administrators and army personnel working for the colonial service in the Indian subcontinent during the nineteenth century dissolved their bitter-tasting antimalarial treatment or tonic (quinine) in gin (a solution of ethanol in water). Henceforward, gin and tonic became established as a fashionable and well-liked alcoholic drink. In the English speaking world, tonic water is appreciated as a mixer for cocktails especially those based on gin or vodka.

To this day, the bitter taste of quinine is used in dilute solution to produce what is known and sold as "tonic water." Quinine, or a quinine salt such as quinine hydrochloride, is dissolved in a small quantity in citrus drinks in order to enhance the flavor of soft drinks such as bitter lemon and bitter lime.

Artemisinin

More recently, a new and completely different antimalarial miracle drug, artemisinin, has been extracted from the leaves of *Artemisia annua* which grows in China and is shown in Figure 2.15. The discovery led to the award of the Nobel Prize in 2015 (see Historical Note).

HISTORICAL NOTE

The plant, *Artemisia annua*, has been used by Chinese herbalists for over two millennia. An extract was believed to have been used in the treatment of skin diseases and malaria. The antimalarial property of the extract was first specifically described in the fourth century within the classic Chinese text, *The Handbook of Prescriptions for Emergencies.*

In the 1960s, a research program was set up by the Chinese army to find an adequate treatment for malaria. By 1972, artemisinin had been discovered in the leaves of *Artemisia annua.* Screening of over 5,000 traditional plants (TCMs) revealed that artemisinin was the most effective drug in dealing with malaria parasites in a patient. Owing to the secret nature of the research program, however, the work was not given until recently the full international recognition it deserved. Today the world is a grateful and beneficial recipient of chemical derivatives based on artemisinin. A share of the 2015 Nobel Prize for Medicine or Physiology was awarded to the Chinese scientist, Youyou Tu for her discovery of the anti-malarial properties of artemisinin.

Artemisinin (Figure 2.16) has a complex structure including a six-membered lactone ring (for more on lactones see the section titled "A Steroid in Your Garden." The structure also contains an unusual peroxide linkage, which is believed to be involved in the antimalarial effectiveness of the drug and may also account for its relatively rapid medical action compared to quinine. WHO recognizes that artemisinin is very effective in the prevention and treatment of malaria even in cases where the parasite responsible is resistant to quinine. To avoid the possibility of drug resistance to artemisinin treatment, WHO recommends a combination therapy of artemisinin and quinine derivatives, respectively.

Carbon–Oxygen Single Bonds and Oxygen–Oxygen Single Bonds in Organic Compounds

The electronegativity of carbon and oxygen atoms is similar arising from the fact that they are close to each other in atomic structure and position in the periodic table.

FIGURE 2.15 **(See color insert.)** *Artemisia annua* (annual wormwood). (Taken from R. Cooper and G. Nicola. 2014. *Natural Products Chemistry: Sources, Separations and Structures*. CRC Press, Taylor & Francis Group.)

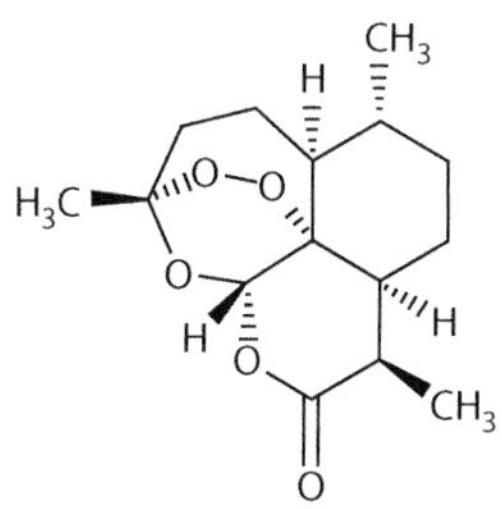

FIGURE 2.16 Molecular structure of artemisinin.

Consequently, carbon–oxygen single bonds are covalent and strong with little charge separation or dipole effect.

In contrast, the peroxide link formed by a single covalent bond between two oxygen atoms, O–O, is relatively weak by comparison with a single C–O bond. We all know that chains, however long, are only as strong as the weakest link because that is the point where a break is most likely to occur. Physically and chemically, the molecules of peroxide compounds are unstable, since they are likely to break at the weakest point, the single O–O bond, resulting in the release of free radicals (see Glossary), which are highly reactive and destructive to tissues in the human body.

The instability of organic peroxide compounds may be further contrasted with the great stability of ethers where the strength of the single C–O–C covalent bonds results in a family of substances, which is relatively stable with little chemistry. These compounds are often volatile liquids at ambient temperature and pressure due to an absence of hydrogen bonding between adjacent molecules. For more on ethers, see the section on "Maca from the High Andes in South America" in Chapter 4.

Furthermore, the stability of C–O single bonds should not be confused with the reactivity of C=O double bonds, which are somewhat strained physically, and weakly dipolar, leading to the extensive chemistry of the large families of organic compounds, which contain the C=O double bond; namely, aldehydes, carboxylic acids, esters and ketones. This book provides many references to the diverse chemistry of each of these families of organic compounds, which are covered extensively in the sections on "Asian Staple: Rice," "A Plant from the East Indies, Camphor," "A Steroid in Your Garden," "Morphine: A Two-Edged Sword" and "Europe Solves a Headache."

The Peroxide Link and the Peroxide Bridge in Artemisinin

A peroxide is a compound containing the link of two oxygen atoms joined together by a single covalent bond as in hydrogen peroxide, H_2O_2, or in a generalized organic peroxide, R′.O.O.R″. The peroxide bond is weak, which renders peroxide compounds unstable, hence they readily decompose to form highly reactive free radicals, R′O and R″O. Owing to this instability, hydrogen peroxide and organic peroxides are found in strictly limited quantities in the natural environment.

Owing to their property as a bleaching agent, peroxide compounds are used in detergents and in cosmetic colorant treatments for human hair. In the laboratory and in some industrial pharmaceutical processes, organic peroxides are utilized to good effect as intermediates or building blocks in step-wise and lengthy preparations.

It is interesting to note that the firefly produces a small amount of the peroxide, 1,2-dioxetane (Figure 2.17), which decays spontaneously to produce electronically excited molecules of acetaldehyde. As each molecule returns to the ground electronic state, a quantum of visible light is released—an instance of chemiluminescence. Quaint glow sticks produced for human entertainment have 1,2-dioxethanedione, C_2O_4, embedded within them, which slowly breaks down in a similar way to yield carbon dioxide and emission of light.

A more sobering consideration is the care required in the storage of liquid organic compounds in the laboratory—particularly ethers. When ether is stored in a Winchester bottle (see Glossary) with some air inevitably enclosed and in the presence of light, unstable peroxides will form very slowly over a period of time measured in months, rendering the bottle somewhat explosive and hazardous to use. This effect is counteracted simply by keeping ether over potassium hydroxide which destroys the peroxide impurity as it develops.

FIGURE 2.17 1,2-Dioxetane also called 1,2-dioxacyclobutane.

QUESTIONS

1. Explain the chemical properties of quinine by reference to its molecular structure.
2. Give an account of the behavior of the peroxide link seen in artemisinin, and relate this to the properties of free radicals in organic chemistry.
3. Explain why peroxides are so unstable.
4. Potassium hydroxide is added to ethers when they are stored in order to avoid peroxide formation. Can you explain how this precaution avoids the risk of accumulation of dangerously explosive peroxides?
5. Give an account of the different chemistry of each of the three C–O bonds in artemisinin.
6. Give a full account of the implications of the high reactivity of introduced hydrocarbons with naturally occurring ozone in the stratosphere of the Earth's atmosphere. Explain what international measures have been undertaken to reduce the release of hydrocarbons to counteract depletion of the ozone layer.
7. Ozone can appear as a pollutant at low levels in the Earth's atmosphere (or troposphere) in the presence of strong sunlight resulting in a photochemical smog, which is injurious to human health. Explain carefully the chemistry, which gives rise to this phenomenon in developed countries of the world.

REFERENCES

J. Jaramillo-Arango. 1949. A critical review of the basic facts in the history of cinchona. *Journal of the Linnean Society of Botany* 351:272–309.

G. Stork et al. 2001. The first stereo-selective total synthesis of quinine. *Journal of the American Chemical Society* 123(14):3239–3242.

R. B. Woodward and W. E. Doering. 1944. The total synthesis of quinine. *Journal of the American Chemical Society* 66(5):849.

SUGGESTED FURTHER READING

R. Cooper and G. Nicola. 2014. *Natural Products Chemistry: Sources, Separations and Structures*. CRC Press, Taylor & Francis Group.

H. Hobhouse. 2015. *Seeds of Change. Six Plants that Transformed Mankind*, Pages 3–50. Counterpoint.

A STEROID IN YOUR GARDEN

Abstract: There is a very important medicine growing in your garden! In Western medicine, the drug known as digitalis has been exploited for its medicinal qualities for many years. Digitalis is obtained from the roots and seeds of the foxglove. In fact, the foxglove contains several important but highly toxic, chemically related steroidal glycosides.

Natural products chemistry

- Cyclic esters—known as lactones
- Lactones as building blocks in nature.

Curriculum content

- The hydroxyl functional group in alcohols (primary, secondary, and tertiary)
- The carboxy functional group in esters.

The Foxglove and Digoxigenin

The foxglove (Figure 2.18) (Latin name: *Digitalis purpurea*) is a wild plant, a native of the woodland margin in temperate climes, which is often cultivated in domestic gardens for its ornamental value.

Foxglove, including the roots and seeds, contains chemicals known as cardiac glycosides. One of these chemicals has the common name digitalis or its chemical

FIGURE 2.18 **(See color insert.)** Foxglove (*D. purpurea*). (Taken from R. Cooper and G. Nicola. 2014. *Natural Products Chemistry: Sources, Separations and Structures*. CRC Press, Taylor & Francis Group.)

name, digoxigenin (Figure 2.19). This compound is composed of two distinctive building blocks, which occur frequently in the natural world; the carbon skeleton of a steroid (see the section on "Central America's Humble Potato!" in Chapter 2) and that of the five-membered cyclic ester (RCOOR) known as a lactone represented in Figures 2.20 through 2.22, as examples.

Digitalis is an example of a drug derived from a plant used by folklorists and herbalists although it is difficult today to determine what amounts of active drug were present in those early herbal preparations. An extract of digitalis was used for the first time in 1785 as the modern era of therapeutic science was beginning. William Withering, a Fellow of the Royal Society, was a physician actively engaged at a hospital in Birmingham, England. He published "An Account of the Foxglove

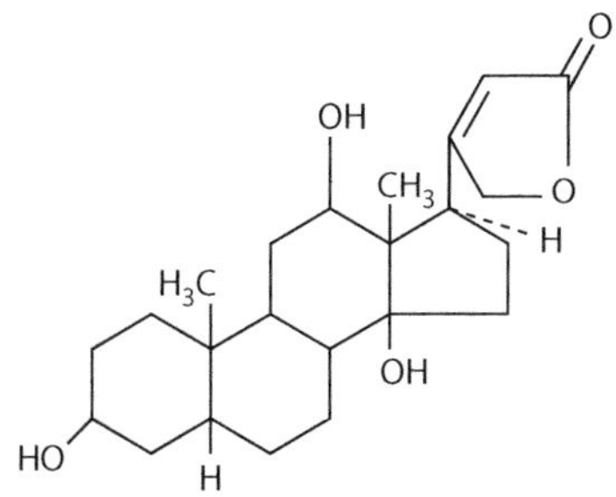

FIGURE 2.19 Molecular structure of digoxigenin.

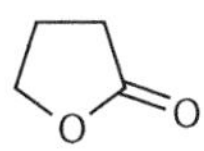

FIGURE 2.20 The structure of 4-hydroxybutyric acid lactone, also known as γ-hydroxybutyric acid lactone.

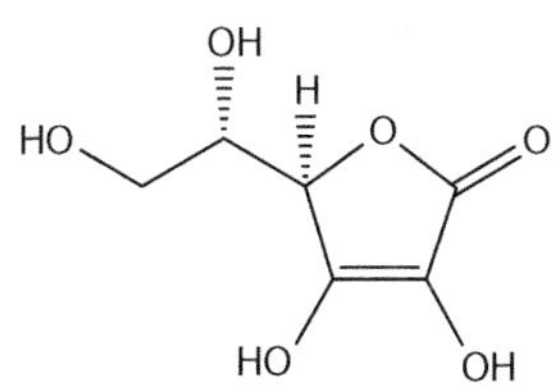

FIGURE 2.21 The structure of vitamin C, also known as ascorbic acid.

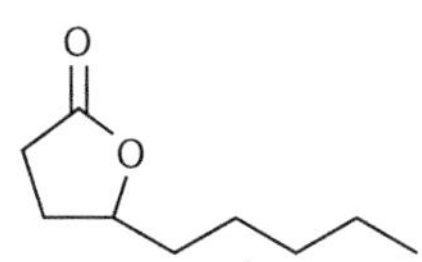

FIGURE 2.22 Chemical structure of γ-nonalactone.

and Some of Its Medicinal Uses," which contained notes on the medical effects of digitalis on congestive heart failure, characterized by low heart output. He had learned of the folk remedies used by people in Shropshire, UK, where he had grown up. Digitalis is used to control a slow heart rate and to strengthen the contraction of heart muscles, thereby producing a stronger pulse. However, digitalis is very toxic and must be administered carefully. An overdose can easily be fatal. It can still be prescribed for those patients who suffer from atrial fibrillation (an erratic heartbeat), especially if they also have congestive heart problems.

A modern application of digoxigenin is in the field of analysis called immunohistochemical staining (IHS). The technique is widely employed in molecular biology to visually reveal the presence of antigens causing abnormality in cells, as in cancerous tumors, and was first reported in the scientific literature by Coons et al. (1941). IHS has proved to be an excellent means of detecting the protein belonging to an antigen within body tissue—especially in neuroscience, where tumors may be located precisely within nerve tissue and brain cells.

Esters Are Formed from Alcohols and Acids

The alcohols are a class of organic substances forming a homologous series with the general formula $C_nH_{2n+1}OH$ containing the characteristic functional group: the hydroxyl. They are rarely found in a free state in nature although a little may be present in over-ripe fruit. The individual names of alcohols are derived by adding the suffix, ol, to the name of the corresponding alkane, thus: methanol, CH_3OH; ethanol, C_2H_5OH; propanol, C_3H_7OH, etc. Where isomerism occurs in propanol and in succeeding members of the series, the position of the hydroxyl group in the longest carbon chain is indicated by inserting a number before the "ol."

Thus, there are three types of alcohol; primary, secondary and tertiary, depending on the chemical environment of the carbon atom to which the hydroxyl is bonded.

$$\text{Primary } R' \cdot CH_2OH$$

$$\text{Secondary } R'R'' \cdot CHOH$$

$$\text{Tertiary } R'R''R''' \cdot COH$$

There is a marked difference between alkanes and alcohols with regard to boiling point and solubility in water. The difference arises unambiguously from the ability of hydroxyl groups to form hydrogen bonds by electrostatic attraction due to their dipolar nature. This molecular association has the effect of increasing molecular weight. Even the alcohols of lowest molecular weight are colorless liquids at normal temperature and pressure. As the alkyl moiety increases in proportion in the molecules of larger members of the series, so the influence of the hydroxyl group declines and physical properties relate more closely to those of the alkanes.

Alcohols undergo two types of chemical reaction directly arising from the hydroxyl group, which may break at the hydrogen–oxygen bond or at the carbon–oxygen bond.

Dehydration of ethanol occurs when the vapor is passed slowly at 200°C over a finely divided catalyst, aluminum oxide. Diethyl ether ($CH_3CH_2 \cdot O \cdot CH_2CH_3$) is the product formed from the breaking of the hydroxyl bond.

$$2C_2H_5OH = CH_3 \cdot CH_2 \cdot O \cdot CH_2 \cdot CH_3 + H_2O$$

However, at a higher temperature of 300°C, ethene is formed as the stronger hydroxyl bond is removed.

$$C_2H_5OH = CH_2CH_2 + H_2O$$

This reaction provides a renewable synthetic pathway in the chemical industry to alkenes and polymers from a feedstock of ethanol produced by the fermentation of glucose obtained from commercially grown plants such as sugarcane.

A primary alcohol may be oxidized through an aldehyde as an intermediate to form a carboxylic acid.

$$R \cdot CH_2 \cdot OH + \text{'O'} = R \cdot CHO + \text{'O'} = R \cdot COOH$$

A typical oxidizing agent for the reaction, which is performed under reflux conditions, is acidified potassium dichromate solution.

Esters are formed by the elimination of water from a reaction between an alcohol and a carboxylic acid, for example:

$$C_3H_7OH + CH_3 \cdot CO \cdot OH = CH_3 \cdot CO \cdot OC_3H_7 + H_2O$$

The reaction is effected by gently refluxing the reagents in the presence of a small quantity of a mineral acid, sulfuric or hydrochloric, as a catalyst.

Properties and Uses of Esters

Esters are formed from alcohols and carboxylic acids, whose properties are fully described in the section titled "Europe Solves a Headache!" and reinforced in "Morphine: A Two-Edged Sword."

The protocol for naming esters parallels that for the naming of salts in inorganic chemistry. Since esters are made from an alcohol and an acid, the alkyl group from the alcohol comes first and the acid stem second.

While esters can be made synthetically in the laboratory, esters occur naturally in many different flowers and fruits. Indeed, ethyl ethanoate smells strongly of the confection known as pear drops.

Esters have a sweet, fruity aroma, which serves well for applications in the perfume industry. Esters are also used as a flavoring agent in the processed food and soft drinks industries.

Owing to the presence of the carboxyl group, esters tend to be polar molecules, so even those of low molecular weight are liquids. These esters find application as a solvent for printing ink where their characteristic smell is noticeable when marker pens are applied by a reader to highlight text on paper.

Esters are also applied in the cosmetics industry where they are used as the medium in which the active ingredients of creams and soothing or healing salve are dissolved or suspended. Furthermore, esters contribute to the aesthetic feel of the product on the skin.

Introducing Cyclic Esters Known as Lactones

The name of the class, lactone, arises from the intramolecular dehydration of lactic acid, $CH_3CH\ (OH) \cdot COOH$. The Latin name for sour milk is *lactis* owing to the presence of lactic acid. A lactone (Figure 2.20) is a cyclic ester which can be formed from an intramolecular reaction as the condensation product of an alcohol group –OH and a carboxylic acid group –COOH. Lactones are characterized by a closed ring consisting of two or more carbon atoms and a single oxygen atom with one of the carbon atoms being part of a ketone functional group, typical of an ester (see the top right of the structure of digoxigenin in Figure 2.19).

Individual lactones are named according to the precursor carboxylic acid

• Aceto	Two carbon atoms
• Propio	Three carbon atoms
• Butyro	Four carbon atoms
• Valero	Five carbon atoms
• Capro	Six carbon atoms

The nomenclature of lactones involves one further refinement. The first carbon atom along the chain of the parent molecule from the carbon atom in the COOH group is labeled alpha (α), the second beta (β) and so on.

Lactone molecules with three or four-membered rings are physically strained and, as a consequence, are quite reactive. In contrast, lactones with five or six-membered rings are relatively straightforward to prepare and are much more stable.

Lactones as Building Blocks in Nature

Lactone rings occur widely in nature as building blocks within larger molecules, which form a part of neurotransmitters, or of various enzymes, or of ascorbic acid also known as vitamin C (Figure 2.21).

Vitamins

The term, vitamin, is derived from "vitamine"—a compound word formed from vital and amine.

Currently, there are 13 recognized vitamins: vitamins A to E, including a range of B vitamins, and vitamin K.

Vitamins fall into two broad categories. The fat-soluble vitamins, vitamins A, D, E and K, can be stored by our bodies in the liver or in fatty tissues. They are stored until they are required, which consequently means they generally do not need to be ingested as frequently. Water-soluble vitamins, on the other hand, are not stored in the body. As such, they must be a regular part of the diet in order to avoid deficiency.

Vitamins are vital for good human health: they are an important part of our diet. They perform a range of roles in the body. For example, a number of the B vitamins are important for making red blood cells and in the metabolism of a variety of compounds during digestion. Others have uses in more specific parts of the body; for example, vitamin A is important for good eyesight, whilst vitamin K plays a major role in the clotting of blood. Conversely, deficiencies of vitamins can also have effects; a lack of vitamin C can lead to scurvy, the bane of sailors before the role of vitamin C was understood. A lack of vitamin K can cause bleeding problems, which is why newborn babies are immediately given a dose containing the vitamin to prevent potential brain damage.

Vitamin C, Ascorbic Acid

Vitamin C is an essential nutrient for sound human health. Vitamin C is a primary metabolite that is directly involved in normal growth, development and reproduction. Also, it plays a role as a redox agent and catalyst in a broad array of biochemical reactions and processes. Vitamin C acts as a reducing agent donating electrons to various enzymatic and nonenzymatic reactions.

Vitamin C is found in fresh vegetables and fruits, and is also present in animal organs such as the liver, kidney and brain. Owing to its antioxidant properties, vitamin C has been widely used as a food additive to prevent or limit oxidation.

HISTORICAL NOTES ON VITAMIN C

A work published in 1753 suggested that citrus fruits (limes) contained certain compounds that could treat scurvy. Scurvy became an endemic disease between the seventeenth and nineteenth centuries because of insufficient dietary intake of fresh fruit and vegetables. Today, we know that vitamin C has the ability to cure and prevent scurvy.

The discovery of vitamin C began in the late sixteenth century when French explorers were saved from effects of scurvy by drinking a tea made from the arbor tree during long sea voyages. Later, it was noted that lemon juice could prevent people from getting scurvy. By 1734, it had been concluded that people who did not eat fresh vegetables and greens would get the disease—a clear risk for sailors denied access due to a long sea voyage in the days of sail. All seamen were thus provided with citrus fruits. The famous British naval captain, James Cook, routinely supplied his men with limes during long voyages of exploration and hydrographic survey in the late eighteenth century. In fact, the practice was commonplace in the Royal Navy of the time so much so that the American term for the British, "limeys," arises from it. These observations led to important breakthroughs in the understanding of scurvy through experiments involving guinea pigs and became one of the first examples of the use of animal models to study disease. The first chemists to isolate vitamin C, ascorbic acid, were Svirbely and Szent-Gyorgyi for which they received the Nobel Prize for Medicine in 1937. The approach takes advantage of the acidic functional groups present in the molecule since ion-exchange resins are used to remove the acidic cations of vitamin C from aqueous solution. When a dilute solution of a strong acid is eluted through the resin as a second step, the anions of the strong acid are retained and vitamin C or ascorbic acid is released.

The first synthesis of vitamin C was achieved by Haworth and Hirst, and also resulted in the award of the Nobel Prize for Chemistry in 1937. Mass production of vitamin C by the Swiss pharmaceutical giant, Hoffmann-La Roche, came 20 years later.

Commercial Uses of Lactones

As is the case with chain or branched chain esters, some lactones, since they are cyclic esters, are used for flavoring processed foods and drinks and as fragrances particularly when specific aromas are required. For instance, γ-nonalactone, presented in Figure 2.22, smells of coconut.

Artemisinin, a complex lactone, is employed in the prevention and treatment of malaria—see the section on "Attacking Malaria: A South American Treasure and a Chinese Miracle."

Gibberellins are another group of complex diterpenoid molecules, some of which have a lactone bridge completing one of the four rings. Gibberellins are worthy of mention, because they are plant hormones, which regulate growth by stimulating plants to grow tall (or to remain short in the absence of them). They also regulate the flowering and ripening of fruit. Gibberellins continue to be the subject of intense research interest due to their ability to increase crop yields.

QUESTIONS

1. Name the following alcohols; $CH_3CH(OH)\cdot CH_3$; $CH_3\cdot CH_2\cdot CH_2\cdot CH_2OH$; $CH_3\cdot CH_3\cdot CH\cdot CH_2OH$; $CH_3\cdot CH(OH)\cdot CH_2\cdot CH_3$; $CH_3\cdot CH_3\cdot CH_3\cdot COH$. Identify the primary, secondary and tertiary alcohols.
2. Give an account of the uses of different alcohols, giving emphasis to ethanol, and their value as feedstock for a variety of industrial processes.
3. Compare and contrast the oxidation of primary, secondary and tertiary alcohols.
4. Given knowledge of the chemistry of esters, describe and give examples of the kinds of reactions lactones will undergo.
5. Three- and four-membered ring lactones are quite reactive, whereas lactones with five- or six-membered rings are much more stable. What is the reason for this in physical terms?
6. What is a vitamin?
7. Give the structure of a gibberellin and point out the important functional groups in the molecule.

REFERENCES

A. H. Coons, H. J. Creech, R. N. Jones. 1941. Immunological properties of an antibody containing a fluoroscein group. *Proceedings of the Society for Experimental Biology and Medicine* 47:200–202.

J. MacMillan. 2001. Occurrence of gibberellins in vascular plants, fungi, and bacteria. *Journal of Plant Growth Regulation* 20(4):387–442.

SUGGESTED FURTHER READING

G. F. Ball. 2004. *Vitamins: Their Role in the Human Body*. John Wiley & Sons.

R. Cooper, G. Nicola. 2014. *Natural Products Chemistry: Sources, Separations and Structures*. CRC Press, Taylor & Francis Group.

M. B. Davies, J. Austin, D. A. Partridge. 1991. *Vitamin C: Its Chemistry and Biochemistry*. Royal Society of Chemistry, paperback series.

R. B. Rucker, J. Zempleni, J. Suttie, D. McCormick, Eds. 2007. *Handbook of Vitamins*, 4th edition. Taylor & Francis.

AFRICA'S GIFT TO THE WORLD

Abstract: The Madagascan periwinkle becomes Africa's great gift to the world. A chain of serendipitous events led the Eli Lily Pharmaceutical Company, based in the United States, to isolate the drug, vinblastine, which is still used today in the fight against childhood leukemia.

Natural products chemistry

- Vincristine and vinblastine
- Alkaloids and indoles
- Isolation of chemicals from plants.

Curriculum content

- Aromatic chemistry with a nitrogen atom within an organic ring
- Indole, and alkaloids.

Discovery of the Periwinkle Plant and Its Properties

Periwinkle, also known as *Catharanthus rosea,* is a tropical perennial, often grown as an annual in temperate climates. The African plant was found only in Madagascar (Figure 2.23).

It was first described in the mid-eighteenth century. In 1757, the Madagascan periwinkle was brought to Europe as an ornamental plant and was cultivated in

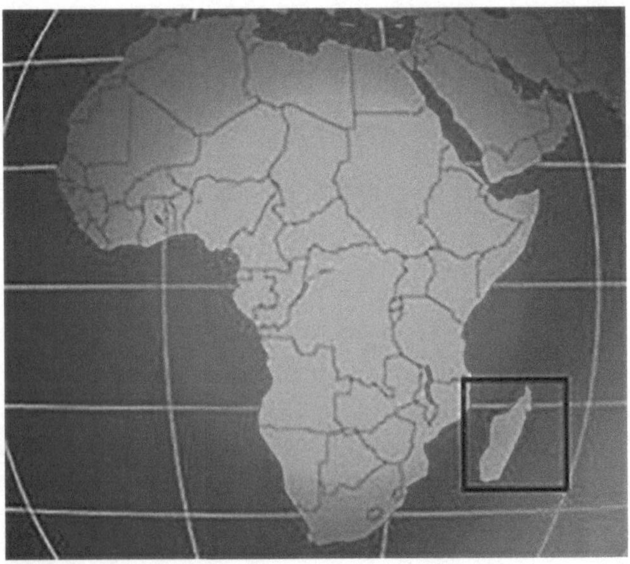

FIGURE 2.23 Madagascar, Island off East Africa. (Courtesy of One World Nations Online, http://www.nationsonline.org/oneworld/madagascar.htm)

European gardens. By the late eighteenth century, widespread distribution throughout the tropics led to its adoption as a medicinal plant wherever it became established

- In Cuba and in Puerto Rico, an infusion of flowers together with a few drops of alcohol (ethanol) added was used as an eyewash for infants.
- In Latin America, the leaf tea has been used as a gargle for sore throat and laryngitis.
- In India, the fresh juice squeezed from the leaves was used for wasp stings.
- In Vietnam, herbalists use the leaf and stem tea as a treatment for everything from menstrual difficulties to malaria.
- In Asian cultures, South Africa and Caribbean islands, periwinkle tea was useful as a folk cure for diabetes.

In the early twentieth century, patent medicines containing periwinkle were touted as a cure for diabetes. "Vinculin" was sold in Great Britain while "Convinca" was sold in South Africa, but both were later shown to have no genuine medical influence in reducing levels of blood sugar. Despite this, consumption of the Madagascan periwinkle as a spurious remedy for diabetes continued and remarkably, led scientists to use the plant as the source for the development of an antileukemia drug.

Indole is an important building block for many naturally occurring alkaloids, which include significant, complex, chemical compounds extracted from the Madagascan periwinkle. Both indole and the indoles as a family of compounds are described in more detail in the sections titled "Maca from the High Andes in South America" in Chapter 4 and "Woad and Indigo" in Chapter 7.

Serendipity

Probably more drugs have been developed through a serendipitous approach by scientists than through planned attack. Educated observation of abnormal events can be further investigated. However, the discovery of the exquisitely complex chemicals from the Madagascan periwinkle was serendipitous. Rather than an inspirational, Archimedean moment of "Eureka," the two active, naturally occurring chemicals extracted from the Madagascan periwinkle were developed from research and careful observation. These compounds were eventually made commercially and sold as drugs. They were called Velban and Oncovin and were developed by a U.S. pharmaceutical company based in Indiana: the Eli Lily Company.

Traditional Madagascan healers used the periwinkle for treating diabetes, which led Western scientists to collect samples of the plant and subsequently to study the plant extract further. Eventually, using sophisticated chemical and biological techniques and some luck, they discovered its anticancer properties quite by accident. Further research led them to isolate and characterize two of the most important chemicals in the plant as cancer-fighting medicines. The chemicals were given the names, vincristine and vinblastine. Today, vinblastine has helped increase the chance of surviving childhood leukemia from 10% to 95%, while vincristine is used to treat Hodgkin's disease. This is a cancer of the lymph tissue, which is found in the liver and in bone marrow and elsewhere, and is expressed in the number of white blood cells.

Thomas Hodgkin described the symptoms in 1832 (see Glossary).

Angela's Story

One medical miracle is the story of little Angela, who, at seven years old, survived childhood leukemia, a form of cancer that up until recent times was almost always fatal. Angela and her family will always be grateful to the doctors who helped her on the road to recovery. As part of her treatment, the doctors prescribed the very special drug derived from the periwinkle called vinblastine, which was central to the chemotherapy that cured her of leukemia. Without the Madagascan periwinkle, she probably would not have survived. Angela endured more than 2 years of medical treatment. Her curly locks of hair fell out, and she remained thin and fragile, yet she bravely went to school. Her class teacher and the class children were so inspired by her courage that they all decided to shave their heads so Angela did not feel different in the classroom. Without vinblastine, her illness could have been fatal and she has every chance of full recovery.

Modern Research and Therapeutic Value

In the 1950s, the scientists, Robert Noble and Charles Beer at the University of Western Ontario in Canada discovered that extracts of the Madagascan periwinkle destroyed white blood cells. However, they had begun research on this plant based on folklore reports of antidiabetic activity from a surgeon, Dr. C. D. Johnson living in Jamaica. The results of their research on the discovery of novel anticancer activity found in the chemical extracts were presented in March 1958 in a research symposium at the New York Academy of Science. Their paper had been submitted at the last minute by invitation of the conference organizer, and was the last of the evening's presentations. The symposium ran late and the Canadians presented their findings at midnight! The audience had by then dwindled to just a few listeners—mostly members of a team of researchers from the Eli Lily drug company.

During this period, Eli Lily was testing and screening hundreds of plant extracts each year in search of biological activity which might lead to the development of a new drug. Natural products chemist, Dr. Gordon H. Svoboda (1922–1994) at Eli Lily, had added the Madagascan periwinkle to the list of research subjects based on reports of use of periwinkle products for the treatment of diabetes in the Philippines during the Second World War. Independently of the Canadians, an extract of the plant was submitted for assay, and Svoboda learned in early 1958 that the extract exhibited very high potency in anticancer tests. So, at the time that the Canadians' paper was presented in the spring of 1958, neither the Canadian research group nor the Eli Lily researchers knew of each other's work on the same plant. However, both groups had observed that their respective plant extracts lowered white blood cell counts in laboratory animals whilst they were looking for antidiabetic effects. Since leukemia involves a proliferation of white blood cells, both teams made observations leading to the deduction that an agent which reduced the number of white blood cells might have potential value in the treatment of leukemia. From an extract of the plant, another scientist, Charles Beer, isolated one specific chemical compound. He named the active compound, vincristine, which was also reported in a scientific paper in

1958 at a cancer research symposium. Needless to say, the Eli Lily research team was extremely interested in this Canadian researcher's work on the Madagascan periwinkle.

After an initial meeting, agreement on collaboration between the Eli Lily Company and the Canadian researchers was secured.

Thus, the race was on!

In March 1961, vincristine was approved by the United States FDA as a chemotherapeutic agent for the treatment of Hodgkin's disease. Even more importantly, a second chemical compound, vinblastine, was isolated by Dr. Svoboda from the Madagascan periwinkle. In July 1963, this drug was approved in the United States for the treatment of childhood leukemia.

Within 2 years of the discovery of the compounds and their antitumor activity and approval as new drugs, Eli Lily had to secure and develop a significant supply of Madagascan periwinkle in order to make production commercially viable. This meant rapid development of farming operations. The extraction of just 1 g of vinblastine from the Madagascan periwinkle required 2,000 lbs of dried leaves. It should be noted that the chemical structures of these compounds are complex. Laboratory synthesis is tedious and expensive, so natural supplies remain the best source.

Tens of thousands of cancer patients, especially those suffering with leukemia and lymphomas, have benefited from the drugs derived from this remarkable medicinal plant, the Madagascan periwinkle, better known to many in tropical climates as a weed, and to the American gardener, as an easy-to-grow ornamental. Serendipity led to an educated guess, and the hunch turned out to be correct.

The Alkaloids, Vincristine and Vinblastine

Although extracts of the Madagascan periwinkle had been used as a folk remedy for centuries, scientific studies were only carried out in the 1950s. Research revealed that extracts from the plant contained up to 70 different alkaloids, many of which are active on metabolic systems in humans. Alkaloids are very common natural compounds produced by a large variety of organisms including bacteria, fungi, green plants and animals. More examples of alkaloids are to be found in the sections on "Morphine: A Two-Edged Sword," "Coffee, Wake Up and Smell the Aroma" and "Cocaine."

Two alkaloids of particular interest isolated from the Madagascan periwinkle were vincristine and vinblastine. The alkaloids are very similar to one another as their chemical structures are very closely aligned (see Figure 2.24).

Vincristine has the empirical formula $C_{46}H_{56}N_4O_{10}$. Vincristine contains many functional groups familiar to the student chemist: carbonyl groups, hydroxyl groups, carboxylic groups, aromatic rings and an indole ring. Despite this, structurally and chemically, vincristine and other alkaloids are extremely complex compounds.

Vincristine and vinblastine are so similar that they differ by only one carbonyl group, present in an aldehyde functional group in the former, which is to be compared with a methyl group in the corresponding location in the latter. Can you spot the difference in Figure 2.24?

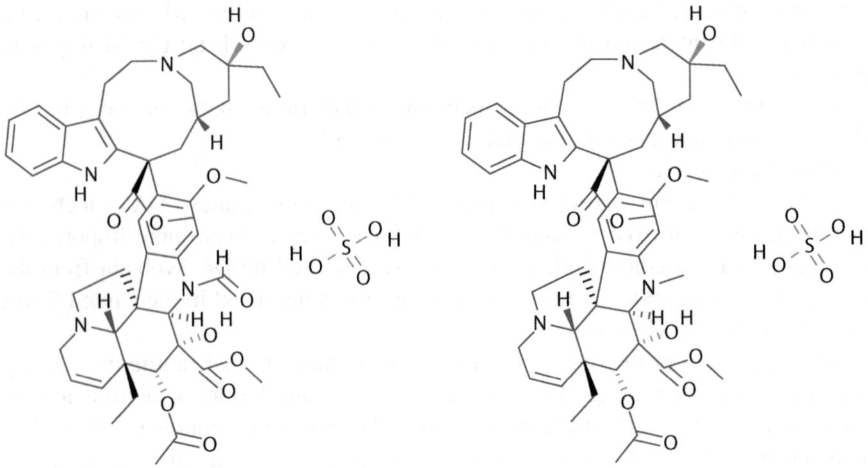

FIGURE 2.24 Vincristine (left) and vinblastine (right) in the form of the sulfate salt. (PubMed. http://pubchem.ncbi.nlm.nih.gov/compound/Vincristine_sulfate)

Furthermore, it should be noted that indole is an important building block for many naturally occurring alkaloids which include significant, complex chemical compounds extracted from the Madagascan periwinkle. Both indole and the indoles as a family of compounds are described in the section titled "Woad and Indigo," in Chapter 7.

Isolation of Vincristine and Vinblastine from Plants

Fractional Distillation

In order to isolate vincristine and vinblastine from the source plant, much research had to be undertaken to find solvents which could dissolve these alkaloids. Thereafter, the first step in the process of extraction involves simply immersing the plant in the solvent. After filtration, the spent plant material is discarded and the solution retained. Then, the solution is concentrated by reducing the volume of solvent which is achieved by fractional distillation. Distillation of the solution leads to the collection of a fraction, which is finally refined by chromatography to yield a pure sample of vincristine.

Chromatography is treated in the section on "An Asian Staple: Rice," while fractional distillation is described in the section titled "Central America's Humble Potato." Furthermore, the process of steam distillation is described in the section on "European Lavender."

Acid–Base Extraction

Alkaloids almost always contain at least one basic nitrogen atom so they can also be purified from crude extracts by acid–base extraction which is a process elaborated upon further in the sections dealing with the decaffeination of coffee and zwitterions, "Coffee: Wake Up and Smell the Aroma" and in "Maca from the High Andes in South America."

QUESTIONS

1. What property allows mixtures to be separated by fractional distillation
   - Density
   - Boiling point or
   - Type of bonding?
2. Explain hydrogen bonding in water and organic liquids, and how it influences density and boiling point.
3. Whenever possible, why is melting point rather than boiling point used to check the purity of a sample of an organic substance?
4. From the structure of vinblastine, point out the indole subunits.
5. Explain simply, in terms of functional groups, the type of reactions you would expect vinblastine to undergo. Draw attention to any complications, which may arise from the polyfunctional nature of the molecule.
6. Give examples of other organic compounds, which have the indole group as a building block, and briefly describe their properties and commercial value.
7. Unlike most amines, indole is weakly basic as only very strong acids such as hydrochloric acid are able to protonate the nitrogen atom. Explain why indole differs from amines in this way.
8. Explain why the acid–base extraction is a useful technique in the purification of vinblastine and vincristine?

REFERENCE

J. Duffin. 2002. Poisoning the spindle: Serendipity and discovery of the anti-tumor properties of the Vinca alkaloids. *Part 1: Pharmacy in History* 44(2):64–76 and *Part 2: Pharmacy in History* 44(3):105–119.

SUGGESTED FURTHER READING

M. Iwu. 2014. *Handbook of African Medicinal Plants*, 2nd edition. CRC Press.

A. D. Osseo-Asare. 2014. *Bitter Roots. The Search for Healing Plants in Africa*. Pages 31–52. University of Chicago Press.

SAVING THE PACIFIC YEW TREE

Abstract: The amazing story of a natural product, paclitaxel (commercially known as Taxol®), which was isolated from a yew tree in the 1960s as an anticancer agent. The development of Taxol was allowed to languish for years due to the issue of a lack of sustainable natural sources. Eventually, the full force of a joint effort by industry and government led to the successful commercialization of the Pacific yew and production of the drug through a process involving fermentation and semisynthesis.

Natural products chemistry

- Terpenes in nature
- Paclitaxel (Taxol).

Curriculum content

- Isomers
- Stereochemistry and chirality
- Nuclear magnetic resonance (NMR) spectroscopy.

The Nature of Paclitaxel

Paclitaxel (Figure 2.25a) is a terpenoid which was first isolated from the bark of Pacific yew trees (*Taxus brevifolia).* Terpenes are hydrocarbons. Terpenes, which contain additional functional groups are known as terpenoids (see also the section on "A Plant from the East Indies, Camphor" in Chapter 6).

HISTORICAL PERSPECTIVE

The compound, Taxol, was discovered in 1971 by Monroe and Wall who were seeking anticancer agents. Remarkably, many more years were to pass before further study at the National Cancer Institute (NCI) in the USA proceeded to clinical trials. The NCI showed great reluctance to pursue Taxol because its isolation was extremely difficult. The bark of the yew tree produced only small amounts of compound and, once stripped of its bark, the tree dies.

However, the program gained momentum due to efforts of Dr. Matthew Suffness at NCI who found Taxol to be very active against melanoma. In 1978, it was also revealed that Taxol had the capability to cause considerable regression in mammary tumors.

In the early 1980s, NCI approached industry for support and Bristol Myers Squibb decided to develop Taxol into a clinically tested drug. New research showed that Taxol could be obtained from the needles of the yew tree—ecologically much better than extraction from bark. Subsequently, Taxol was discovered as a fungal metabolite, which provided the opportunity for large-scale production from fermentation.

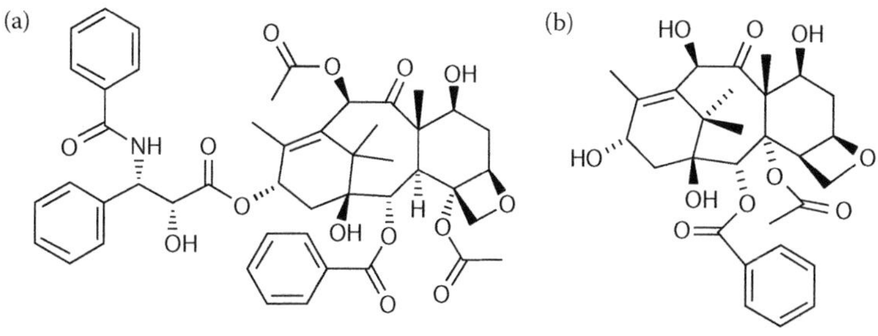

FIGURE 2.25 (a) Paclitaxel and (b) 10-DAB.

Medical Value of Paclitaxel

Paclitaxel was approved by the FDA for treatment of drug-resistant ovarian and breast cancers, and is also used in the treatment of lung cancer. Subsequently, it has become a major research tool of study in cancer therapy. Paclitaxel works by inhibiting mitosis: as the cells are unable to multiply, tumors are unable to grow.

Modern-Day Preparation of Paclitaxel

Today, the preparation of paclitaxel involves an elegant combination of measures. Large amounts of a precursor compound are isolated, which is followed by an additional semisynthetic step to yield the final product. The source of the precursor is the English yew, *Taxus baccata.* The precursor, 10-deacetyl baccatin (10-DAB) (Figure 2.25b), is available from the needles of the tree in large quantity. The difference between 10-DAB and paclitaxel is simply that 10-DAB has no ester side chain. The prepared side chain is attached to the C-13 hydroxyl group of 10-DAB to obtain paclitaxel on a large scale. In this manner, paclitaxel is manufactured by semisynthetic production from a natural precursor.

Terpenes and Isoprene as a Building Block in Nature

Terpenes are a class of compounds which are common in nature. They are also described in the section on "Cannabis and Marijuana" in Chapter 5, and in the section on "A Plant from the East Indies, Camphor" in Chapter 6.

Terpenes are present in tree resin from which turpentine can be extracted. Indeed, the very name, terpene, is derived from the word turpentine. Terpenes are major biochemical building blocks within nearly every living creature. Steroids, for example, are derivatives of the terpene known as squalene. Terpenes and terpenoids are also the primary constituents of the oils of many types of plants and flowers. These oils are used widely as natural flavor additives for food, as fragrances in perfumery, and in both traditional and alternative medicines such as aromatherapy (see the section concerning "Exotic Potions, Lotions and Oils" in Chapter 6).

Terpenes are polymeric compounds made up of a number of isoprene molecules, namely 2-methyl-1,3-butadiene, with a molecular formula $CH_2C\ (CH_3) \cdot CH \cdot CH_2$, which have a short five-carbon chain. Because all terpenes share a common building block, isoprene, terpenes can be categorized by the number of isoprene units they include. The simplest molecule is a two-isoprene unit called a monoterpene which has 10 carbon atoms. Examples of derivatives of monoterpenes are camphor and menthol, which help clear mucus from sinuses when a person is suffering from a cold, and pinene, which is used in wood varnish.

Isomers

Linkage between two isoprene molecules can occur in different ways involving either the "head" or "tail" of the molecule to form three distinct structural sequences, which have the same empirical formula. As a consequence, each of these structurally different molecules has distinct chemistry. Each structurally different molecule is known as an isomer.

Here is an isomer being formed from the head-to-head or 1–1 link

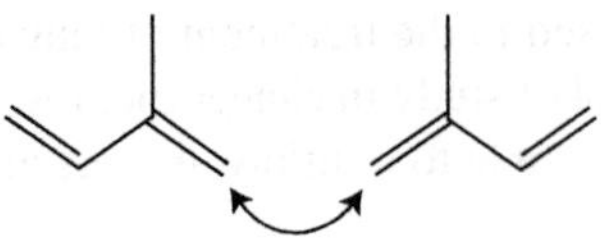

while this link is called a head-to-tail or 1–4 link

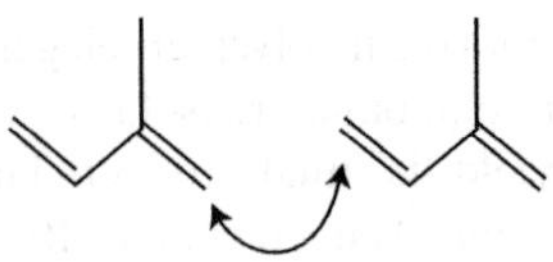

A rare linkage is called a tail-to-tail or 4–4 link

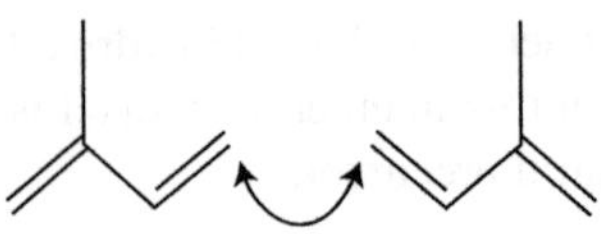

Isomers and Stereochemistry

Physical rotation around the axis of a carbon–carbon double bond is not physically possible, as the double bond is structurally rigid. This gives rise to stereochemistry because the relative orientation of the groups within a molecule can also give rise to different isomers.

trans-1,2-dichloroethene *cis*-1,2-dichloroethene

FIGURE 2.26 An example of two isomers: *trans* (opposite) and *cis* (same). (Taken from R. Cooper and G. Nicola. 2014. *Natural Products Chemistry: Sources, Separations and Structures*. CRC Press, Taylor & Francis Group.)

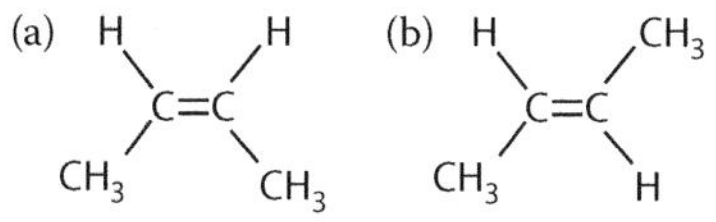

FIGURE 2.27 Stereoisomerism in alkenes. (a) *cis*-2-Butene and (b) *trans*-2-butene.

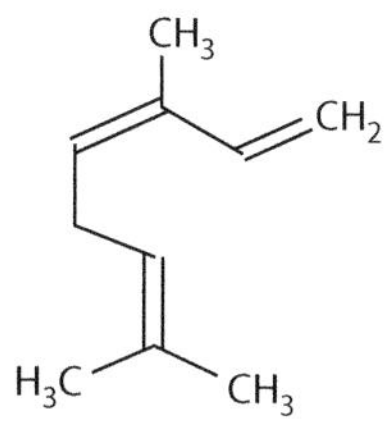

FIGURE 2.28 Structure of ocimene.

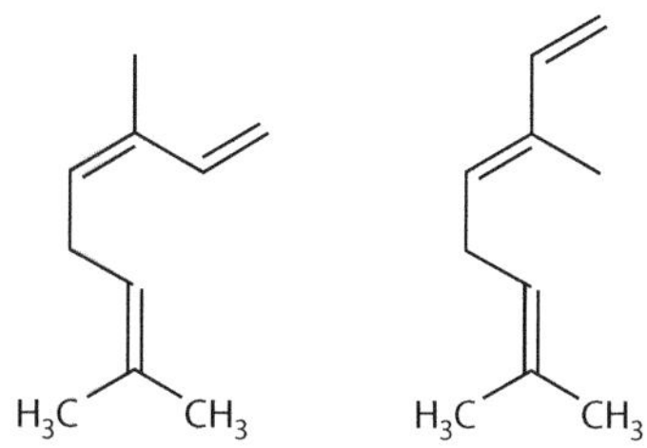

FIGURE 2.29 The geometric isomers of ocimene.

The stereochemistry of carbon–carbon double bonds is usually called *cis* and *trans* isomerism, but is also known as geometric isomerism (Figure 2.26). The terms *cis* means "on the same side" and *trans* means "across."

An alternative, the E–Z notation, makes it possible to deal with more complex cases. Look at the atoms or groups attached to each of the carbon atoms in the double bond in the example below for E- and Z-butene (Figure 2.27). When the two methyl groups are on the same side of the C=C, the isomer is described as Z, from the German word for together, *zusammen*. If not it is E, from the German word for opposite, *entgegen*.

Ocimene (Figure 2.28) is another example of a compound, which has geometric isomers found in the *cis* and *trans* configurations (Figure 2.29). It is extracted as an essential oil from the popular herb, basil (*Ocimum tenuiflorum*). Ocimene is classified as a linear terpene and possesses a pleasant odor and so finds application in perfumery.

Isomers and Chirality

The term chiral is used, in general, to describe an object, which is not superimposable on its mirror image. Human hands are an example. No matter how our two hands are orientated, it is impossible for all the major features of both hands to coincide in space. This difference in symmetry becomes obvious if a left-handed glove is placed on a right-handed glove.

The concept of chirality is extremely important in chemistry too. Each mirror image of a chiral molecule is a special type of isomer known as an enantiomer. A pair of enantiomers is often designated as "right-handed" and "left-handed." Molecular chirality is of considerable interest in stereochemistry, especially in organic chemistry and biochemistry, where molecules can be large and complex. Spatial relationships between functional groups in different, large or complex molecules can fundamentally determine vital aspects of chemical interaction (see also the fascinating entry on the significance of the chirality of proteins and enzymes in the section on "Foods of the Fertile Crescent: Ancient Wheat" in Chapter 3).

A simple example of a chiral molecule arises in a tetrahedral molecule in which all four substituents are different, for example, fluoro chloro bromo methane, $CHFClBr$.

An enantiomer is a chiral molecule which also has the property of rotating the plane of polarized light. If the rotation of light is clockwise (as seen by a viewer toward whom the light is traveling), the enantiomer is labeled (+). Its mirror image is labeled (–). This property helps scientists to study chirality and enantiomers.

Nuclear Magnetic Resonance (NMR) Spectroscopy and Molecular Structure

The impact of NMR spectroscopy on organic chemistry has been substantial. NMR spectroscopy is frequently used by chemists and biochemists to investigate the structure and, hence, the properties of organic molecules from small to large and complex, such as proteins or nucleic acids or carbohydrates.

The technique relies on the phenomenon of NMR. Only nuclei having an odd number of protons or neutrons give a signal. Research often exploits the magnetic properties of the nuclei of the hydrogen atom ^{1}H, usually referred to as a proton in this context, and an isotope of carbon, ^{13}C, which is about 1% abundant compared to the common isotope, ^{12}C.

A spinning charged atomic nucleus generates a tiny magnetic field. When an external magnetic field is applied, the difference between two energy levels (ΔE) is resolved arising from two possible directions of spin. The nuclei of some atoms align with the magnetic field while other nuclei line up against it.

Irradiation of the sample with energy in the radio-frequency band which corresponds to this small difference will cause excitation of those nuclei in the lower energy state (with the field) to the higher energy state. Resonant absorption of energy occurs at a frequency characteristic of the NMR-active atom, ^{1}H or the isotope of carbon, ^{13}C. The spinning nuclei, which lie against the applied magnetic field, will, of course, radiate energy at this frequency when they return to the ground or lower energy state in which the spinning nuclei are aligned with the field.

However, the magnetic field around a nucleus is also influenced slightly by the magnetic effects of the electrons present in other atoms within the molecule and most importantly by the distribution of the other atoms in the molecule. This, of course, is determined by the structure of the molecule. The energy difference between the spin states of the NMR-active nucleus is therefore, affected by the structure, which in turn, alters the exact frequency of absorption of energy by a nucleus in a given externally applied magnetic field strength. Analysis of a spectrum of absorption signals from NMR-active nuclei (such as a proton or the isotope of carbon, ^{13}C) can, in expert hands, reveal details of molecular structure including isomerism (see the section on "Foods of the Fertile Crescent: Ancient Wheat" in Chapter 3, for the application of NMR spectroscopy to the understanding of the structure of proteins and enzymes). As there are fewer sharp absorption peaks in ^{13}C NMR spectra, they are usually more straightforward to interpret than proton spectra. Each functional group in organic chemistry, such as C–C, C=C, C–O, C=O, C–Cl, C–N, CNO, CHO, COOH or aromatic carbon rings, can be identified in ^{13}C NMR spectra from their distinctive chemical shift.

Fine details of the location of hydrogen atoms in a molecule can also be obtained from proton NMR spectroscopy as shown in the spectrum of ethanol below.

The chemical shift of the ^1H nuclei is the difference between its resonant frequency of absorption of energy and the frequency of absorption of proton nuclei in a standard such as tertramethyl silane, which is set at zero on the NMR scale. The chemical shift of each group of protons, whether those present in CH_3, CH_2 or OH in this instance, is influenced by the charged electrons in their immediate proximity, in other words, by the electrons in the functional group of which they are a part of.

The signals from protons in the methyl and methylene groups are, in turn, split into multiple peaks or multiplets due to coupling with the spin of hydrogen atoms on adjacent carbon atoms. Spin–spin coupling is used to interpret NMR proton spectra through what is simply called the n + 1 rule, where n is the number of hydrogen

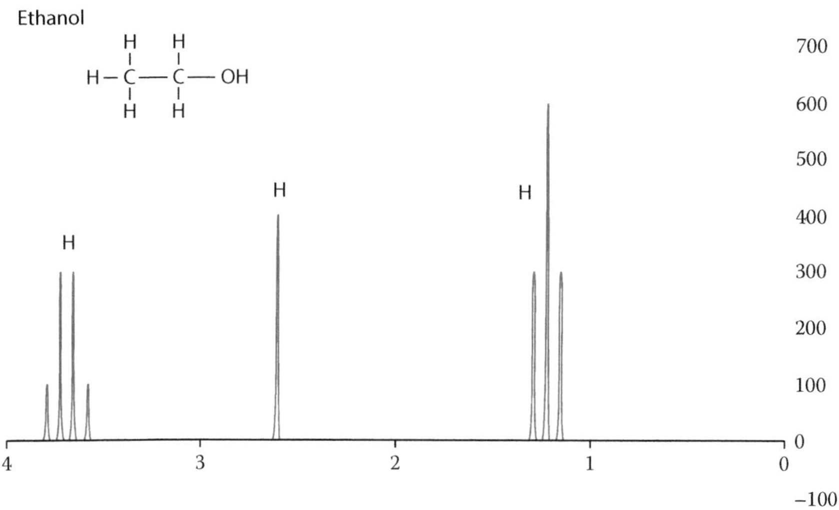

FIGURE 2.30 (See color insert.) ^1H NMR spectrum of ethanol. (Taken from R. Cooper and G. Nicola. 2014. *Natural Products Chemistry: Sources, Separations and Structures.* CRC Press, Taylor & Francis Group.)

atoms on the adjacent carbon atoms. As an illustration in Figure 2.30 we show the ^{1}H NMR spectrum of ethanol.

When a peak in a proton NMR spectrum is split into two multiples, then there is one hydrogen atom on the neighboring carbon atoms.

When a peak in a proton NMR spectrum is split into three multiples, then there are two hydrogen atoms on the neighboring carbon atoms.

When a peak in a proton NMR spectrum is split into four multiples, then there are three hydrogen atoms on the neighboring carbon atoms.

QUESTIONS

1. Although the structure of a molecule of paclitaxel appears complex, its characteristics will be influenced by the functional groups present. Identify those functional groups and their chemical properties.
2. How might an ester side chain be added at C-13, and what complications would you anticipate?
3. Which of the following exhibit(s) geometric isomerism
   - But-2-ene
   - But-2-yne
   - Phthalic acid based on a benzene ring with formula C_6H_4 $(COOH)_2$?
4. Compare and contrast the chemical and physical properties of two structural isomers, namely ethanol and dimethyl ether.
5. Explain why spin–spin coupling is not observed in ^{13}C NMR spectra.
6. Applying the $n + 1$ rule to the multiplets present in the proton NMR spectrum of ethanol shown in Figure 2.30, identify the signals from distinct hydrogen atoms in the CH_3, CH_2 and OH functional groups.
7. Explain in each case why tetramethylsilane, $(CH_3)_4Si$, or deuterated chloroform, $CDCl_3$, may be used as a standard in proton NMR spectroscopy.
8. Give an account of the value of NMR to society in general given its application in approaches to health scanning undertaken in hospitals.

REFERENCES

A seminal paper describing the isolation and structure of the anticancer compound Taxol from the Pacific yew:

M. C. Wani, H. L. Taylor, M. E. Wall, P. Coggon, A. T. McPhail. 1971. Plant antitumor agents. VI. The isolation and structure of Taxol, a novel antileukemic and antitumor agent from *Taxus brevifolia*. *Journal of the American Chemical Society* 93:2325–7.

Key papers describing the chemical synthesis of Taxol:

R. A. Holton et al. 1994. First total synthesis of Taxol. Functionalization of the B ring. *Journal of the American Chemical Society* 116:1597–1598.

K. C. Nicolaou et al. 1994. Total synthesis of Taxol. *Nature* 367(6464):630.

3 Modern Miracles of Foods and Ancient Grains

INTRODUCTION

Over millennia, disparate human populations have learned through an empirical approach to select plants that tasted well, offered nutritional value, were useful in cooking, could be ground up as a spice to improve flavor or could be utilized as a preservative.

In this chapter, we concentrate on the cereals and grains that have sustained civilizations across the globe. Historically, grain agriculture has been successful because the grains can be stored, measured and transported more readily than many other kinds of food crops such as fresh fruits, roots and tubers. An excess of food could be produced and stored easily which contributed to the creation of the first permanent settlements. Also, early civilizations developed techniques in planting and in planned farming in order to maximize yield, which has influenced modern, efficient practice in agronomy.

Much knowledge has been accrued in anecdotal form and passed on orally through generations as traditional wisdom. Some of this knowledge has been found clearly recorded in documents left by major civilizations of the past.

REDISCOVERING TRADITIONAL GRAINS OF THE AMERICAS: CHIA AND QUINOA

Abstract: Chia from Central America and quinoa from South America lead a modern renaissance in appreciation of the value of ancient grains. Chia grains have high levels of desirable fatty acids, which are usually found only in fish. They are not cereal crops, and so the grains are free of gluten.

Natural products chemistry

- Lipids, also known as fatty acids.

Curriculum content

- Alkanes and linear carbon polymers
- Alkenes
- Saturated and unsaturated fatty acids
- Human health issues
- Vegetable oil and biodiesel.

Chia

Chia, *Salvia hispanica*, is a flowering herb. The plant belongs to the mint family and is native to Central America. Chia yields a grain-like seed, which provides a rich vegetable source of omega-3 fatty acids. Chia has been domesticated for the first time in modern history to supply economically viable quantities for world demand. Chia, literally meaning "oily," originated in Mexico and was cultivated by the Aztecs. It is grown commercially today in regions of Central and South America. The useful parts of chia are the seed and the sprout. The seed contains an important lipid, an omega-3 fatty acid known as α-linolenic acid (see Figure 3.1), together with significant concentrations of dietary fiber, protein, calcium, magnesium, iron and antioxidants.

Scientific reports lay emphasis upon the importance of omega-3 fatty acids for our general health. It has been postulated that a deficiency in omega-3 fatty acids in the diet could lead to a greater risk of heart disease, cancer and cognitive defects. The richest dietary source of omega-3 fatty acids is found in fish. However, eating large amounts of fish has been linked with unwanted intakes of heavy metal pollutants, mainly mercury, and possibly other organic pollutants, such as polychlorinated biphenyls. Chia, therefore, offers a reasonable plant alternative.

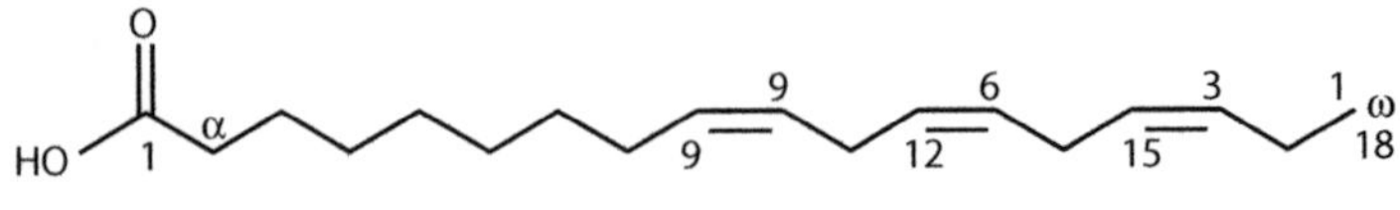

FIGURE 3.1 The structure of the omega fatty acid, α-linolenic acid.

Quinoa

Quinoa is another ancient grain from the plant, *Chenopodium quinoa,* whose value is often overlooked. *C. quinoa* originated in the cooler Andean regions of Ecuador, Bolivia, Colombia and Peru. It was successfully domesticated 3,000–4,000 years ago for human consumption. The Incas, who held the crop to be sacred, referred to quinoa as *chisaya mama* or "mother of all grains." It was the Inca emperor who would traditionally sow the first seeds of the season using "golden implements."

During the European conquest of South America, the Spanish colonists scorned quinoa as "food for Indians," and even actively suppressed its cultivation. The name is derived from the Spanish spelling of the Quechua word *kinwa*. Its nutrient value is very high compared with other modern cereals.

The grains contain essential amino acids. An *essential* amino acid, or *indispensable* amino acid, cannot be synthesized *de novo* (from scratch) by humans, and therefore, must be supplied by the diet. Amino acids are the building blocks for peptides and proteins (see the section on "Foods of the Fertile Crescent: Ancient Wheat" in this chapter).

The grains also contain significant amounts of the minerals calcium, phosphorus and iron. After harvest, the grains need to be processed to remove the coating containing bitter-tasting saponins, and are generally cooked in the same way that rice is prepared.

Quinoa is a grain-like crop grown primarily for its edible seeds. Although quinoa seeds resemble grain in appearance and in characteristics as a food, quinoa is not a member of the grass family and, therefore, is not considered as a true cereal. In fact, quinoa is closely related to species such as beets, spinach and tumbleweeds. Its nutrient values, however, compare well with those of common cereals.

Quinoa was of great nutritional importance in pre-Columbian Andean civilizations, second only to the potato, which was followed in importance by maize. In contemporary times, this crop has become highly appreciated for its nutritional value, as its protein content is very high (18%). Unlike wheat or rice (which are low in lysine), quinoa contains a balanced set of essential amino acids for humans, making it a complete protein source. Today, quinoa seeds may be ground into flour for baking bread or cakes. Quinoa seeds are often simmered gently in water, rather as rice grains are, to make a porridge-like breakfast dish, or may be served in savory salads. Even more importantly, quinoa is gluten free and considered easy to digest, offering an important nutrient for those suffering from gluten-related diseases.

Gluten Sensitivity

Gluten is a mixture of the proteins, gliadin and glutenin, which contribute to the elasticity of dough. Wheat protein and the wheat starch are easily digested by nearly 99% of the human population. However, some research studies in several countries worldwide suggest that approximately 0.5%–1% of these populations may have a genetic disorder, which gives rise to celiac disease. This condition is caused by an adverse reaction of the immune system to gliadin present in gluten. Gliadin is a monomeric protein which is soluble in water, whereas gluten is polymeric and insoluble in water.

Over time, the digestive tract of the small intestine becomes inflamed and serious damage to the lining of the intestinal wall can result. In turn, the ability of the human body to absorb nutrients is affected with resultant discomfort and a general failure to thrive. Hence, susceptible people need a gluten-free diet.

As noted, quinoa is gluten free and is being considered a possible crop in NASA's Controlled Ecological Life Support System for long-duration human-occupied spaceflights.

Lipids

Lipids are a class of organic compounds, which, as fatty acids, are insoluble in water yet soluble in many organic solvents. They include natural oils, waxes and steroids. Lipids, along with carbohydrates and proteins, are important components of all plant and animal cells.

Lipids of medium-chain length are often referred to by common names, which reflect their origin, for instance, palmitic acid, $CH_3(CH_2)_{14}COOH$, comes from the oil of palm grown extensively in the East Indies, whereas stearic acid, $CH_3(CH_2)_{16}COOH$, is a component of soap.

Naturally occurring fatty acids may be saturated (and have no double bonds, like those described above) or unsaturated. Two examples of unsaturated lipids are presented in Figure 3.2. You can see that these molecules are made up from essentially the functional groups of a carboxylic acid, a long-chain alkane and an alkene. These two compounds are geometric isomers. They have the same molecular weight and an identical molecular formula, but there is a difference in the orientation around the double bond known as *cis* and *trans*, as explained earlier (see Figure 2.26).

Esters of fatty acids are classed also as lipids. As a consequence of the size of the molecules and the presence of these well-understood functional groups, lipids exhibit a great deal of structural variety.

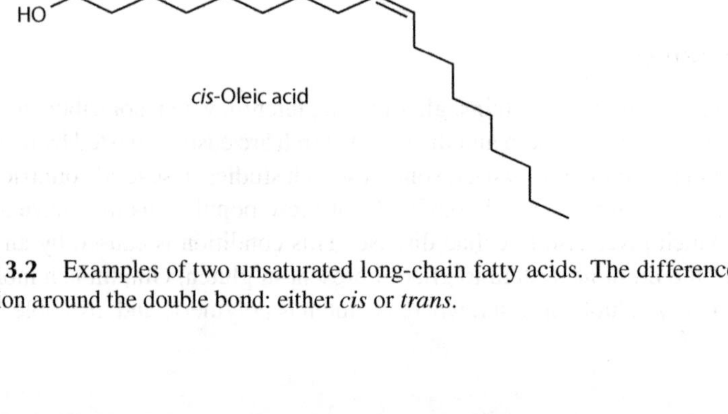

FIGURE 3.2 Examples of two unsaturated long-chain fatty acids. The difference is in the substitution around the double bond: either *cis* or *trans*.

Fats, Oils and Human Health

Fatty acids with three carboxyl groups form triesters with glycerol (propane-1,2,3-triol). Water molecules are formed along with the ester in what is known as a condensation reaction, described further in the next section. These triglyceryl esters (or triglycerides) compose the class of lipids known as fats and oils (for more about esters see the section on the foxglove titled "A Steroid in Your Garden" in Chapter 2). Triglyceryl esters are found in both plants and animals, and form one of the major components of the human diet. Triglycerides, which are solid at room temperature are classified as fats and occur predominantly in animals. Those triglycerides that are liquid are called oils and are found chiefly in plants, although triglycerides from fish are largely oils too. As might be expected, fats have a predominance of saturated fatty acid in the molecule, whereas oils are composed largely of unsaturated acids.

Fats effectively hold a lot of energy in the extensive covalent bonds of the alkane chain, which is why mammals store fats in their tissues. When the body needs energy, fats can be oxidized to release it. Excess sugar in the diet is converted to fat. Low fat and low sugar regimes, coupled with food based on plant products, are of value in a healthy diet, particularly in developed countries of the world, where physical exercise among inhabitants is at a premium.

The animal fat in human food largely consists of saturated triglycerides but it also contains a small amount of cholesterol. In contrast, cholesterol is not found in significant quantity in plant sources, which has ramifications for the human diet, since the ingestion of animal fat is one factor that influences blood cholesterol levels. Cholesterol is produced partly in the human body and is partly absorbed from the animal fats found in eggs, dairy products and meat.

As a complex lipid, cholesterol is only slightly soluble in water and, therefore, it dissolves in the bloodstream only at exceedingly small concentrations and can easily be brought out of solution. An accumulation of material (sometimes called plaque), including cholesterol and fatty acids, can build up on the inner wall of an artery and may narrow the artery thereby restricting blood flow. High blood pressure, greater risk of heart attack and increased susceptibility to heart disease can result.

Plant (vegetable) oils are extensively converted on an industrial scale to solid triglycerides by partial hydrogenation of their unsaturated components. The hydrogenation of vegetable oils to produce semisolid products, which are used in the production of processed foods such as margarine or ice cream, has had unintended consequences. Although hydrogenation imparts desirable features in processed food, when compared with those of naturally occurring liquid vegetable oils, such as spreadability, texture and lengthened shelf life, it introduces some serious health problems. These occur when the *cis* double bonds in the fatty acid chains are not completely saturated in the hydrogenation process. The catalysts used to effect the addition of hydrogen cause isomerization of the remaining double bonds from the *cis* to the *trans* configuration. Unnatural fats in the *trans* configuration appear to be associated with an increased incidence of heart disease, cancer, diabetes and obesity, as well as difficulties in immune response. Because of these concerns, attention has turned to the process of trans-esterification. Natural esters in vegetable oils are converted into different esters with much higher molecular weight, which melt at higher temperatures. These compounds provide the benefit of a smooth consistency to margarine yet avoid the *trans* isomers associated with health issues.

Trans-esterification simply involves reacting an ester with an alcohol to obtain a new ester, that is, the alcohol stem of the original ester is replaced with a different alcohol stem as in the following example:

$$CH_3COOCH_2CH_3 + CH_3CH_2CH_2OH = CH_3COOCH_2CH_2CH_3 + CH_3CH_2OH$$

Ethyl ethanoate combines with propanol in a reaction, which is reversible to produce propyl ethanoate and ethanol. More generally, this reaction can be represented by

$$CH_3COOR + {}^1ROH = CH_3COO^1R + ROH$$

where R and ^{1}R are different stems of alcohols.

Waxes

Waxes are lipids and lipids are esters of fatty acids with long-chain monohydric alcohols (one hydroxyl group). Natural waxes are often mixtures of such esters and may also contain hydrocarbons.

Waxes are widely distributed in nature. The leaves and fruits of many plants have waxy coatings, which may protect them from dehydration and small predators. The feathers of birds and the fur of some animals have similar coatings, which serve as a water repellent.

The chemical formulae of two well-known waxes, bees wax and carnuba wax, are, respectively,

$$CH_3(CH_2)_{24}CO_2-(CH_2)_{29}CH_3 \quad \text{and} \quad CH_3(CH_2)_{30}CO_2-(CH_2)_{33}CH_3$$

Bees wax has many uses, which include the manufacture of fine candles. It is also applied to decorate and protect fine furniture. Carnuba wax is valued for its toughness and water resistance, and so it finds application as a polish or as a sealant.

Vegetable Oil and Biodiesel

Biodiesel is a renewable fuel used in diesel engines. Biodiesel is made from vegetable oil or animal fat. A particular advantage of biodiesel is that it can be produced by recycling vegetable oils of poor quality, or processed from oils, which are left over from human activity, such as cooking oil. Animal fat from food processing plants can also be used.

Essentially, biodiesel is a mixture of the methyl and ethyl esters of fatty acids, which are produced by reacting methanol or ethanol with the triglycerides of oils and fats. This reaction is another example of trans-esterification.

Usually, biodiesel is not used in pure form, but is mixed with conventional diesel, obtained from the cracking of petroleum. As an example, fuel denoted as B20 would contain 20% biodiesel. It should be noted that vegetable oil, as a renewable product, could be obtained from crops, such as the oil seed, rape, grown on a much greater scale than at present, in order to provide a means of reducing human dependency on petroleum, which is not replaceable.

QUESTIONS

1. The richest dietary source of omega-3 fatty acids is fish. However, eating large amounts of fish has been linked with unwanted intakes of heavy metal pollutants such as mercury, (and possibly other pollutants, such as polychlorinated biphenyls). How could trace concentrations of heavy metal pollutants in the open ocean be intensified to significant levels for human health through the natural food chain of fish?
2. Explain why lipids are hydrophobic.
3. In organic chemistry, the determination of melting point is a well-known technique for establishing the purity of a sample dependent on the fact that the melting points of many organic compounds are so precise. Lipids which are saturated acids have higher melting points than lipids of corresponding size which are unsaturated acids. Explain why this is so.
4. Given the functional groups present in saturated lipids and unsaturated lipids, provide different examples of both, and describe the chemical properties which you would expect them to possess.
5. Explain why *trans* fatty acids are considered harmful to human health.
6. Give a balanced account of how reasonable it would be to produce biodiesel on a very large scale from vegetable oils bearing in mind alternative options in agricultural practice. Make sure you also address the greenhouse effect and carbon neutrality in your answer.
7. Figure 3.1 presents an example of an unsaturated fatty acid. What chemistry would you suggest to create the respective saturated molecule? Give the chemical equation.
8. What is the chemical difference between an unsaturated and a saturated fatty acid? What are the implications for human health?

SUGGESTED FURTHER READING

R. Ayerza Jr. and W. Coates. 2005. *Chia*. University of Arizona Press.
J. F. Scheer. 2001. *The Magic of Chia. Revival of an Ancient Wonder Food*. Frog Ltd.

FOODS OF THE FERTILE CRESCENT: ANCIENT WHEAT*

Abstract: Ancient grains, such as *Triticum* originating in the Fertile Crescent at the start of civilization, led to the spread of wheat throughout the world and are related to major contemporary questions concerning population growth, genetic modification and use of fertilizers.

Natural products chemistry

- Phenylpropanoids
- Lignans and lignin
- Germination inhibition.

Curriculum content

- The condensation reaction and hydrolysis
- Amino acids and the peptide link as a building block
- Proteins and enzymes
- Elucidating the molecular structure of proteins
- The shikimic acid pathway in plant metabolism
- Society's challenges in enhancing agricultural production—crop yield and green issues including the genetic modification of crops.

Possible Origins of Ancient Wheat

The word *cereal* derives from *Ceres*, the name of the Roman goddess of harvest and agriculture. Cereals are grasses and are also known as Gramineae. They are cultivated for the edible components of their grain composed of the endosperm, germ and bran.

Archaeological findings indicate that ancient wheat varieties, shown in Figure 3.3, first occurred in parts of Turkey, Lebanon, Syria, Israel, Egypt and Ethiopia. Domesticated einkorn wheat in Turkey dates back to 9,000 BC. Evidence of the existence of wild barley goes as far back as 23,000 BC. Cultivation of wheat began to spread beyond the Fertile Crescent after about 8,000 BC. Jared Diamond in his excellent book, *Guns, Germs and Steel*, traces the spread of cultivated emmer wheat starting in the Fertile Crescent about 8,500 BC, reaching Greece, Cyprus and India by 6,500 BC, Egypt shortly thereafter, followed by introduction in Germany and Spain by 5,000 BC.

The early Egyptians used bread and developed baking into one of the first large-scale food production industries. By 3,000 BC, wheat had reached England and Scandinavia. A millennium later it reached China. Modern wheat has successfully spread globally and is widely cultivated as a cash crop because it produces a good

* Published in part R. Cooper. 2015. Re-discovering ancient wheat varieties as functional foods. *Journal of Traditional and Complementary Medicine*, 5:138–143.

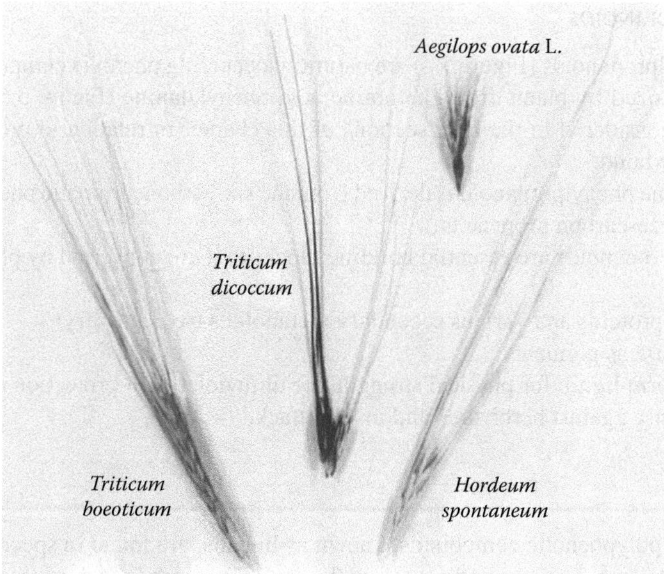

FIGURE 3.3 Ancient varieties of wheat and barley. (Taken from R. Cooper. 2015. *Journal of Traditional and Complementary Medicine*, 5:138–143.)

yield per unit area, grows well in a temperate climate and has a moderately short growing season of about 120 days.

Crop Domestication

Cultivation and repeated harvesting and sowing of the grains of wild grasses led to the creation of domestic strains. Domesticated wheat has larger grains and the seeds (spikelets) remain attached to the ear by a toughened stem or rachis during harvesting. In the case of wild strains, a more fragile rachis allows the ear to easily shatter and disperse the spikelets. As the traits that improve wheat as a food source also involve the loss of the plant's natural seed dispersal mechanisms, highly domesticated strains of wheat cannot survive in the wild.

Crop Germination

A scientific study has been conducted on varieties of ancient wild wheat, particularly *Triticum* and *Aegilops* species, in order to improve understanding of germination of the seeds. Research has concentrated upon the influence of the *amount* of phenolic and flavonoid constituents present. When results were compared with the outcome of similar studies of the phenolic compounds present in 38 species of modern wheat, a consistent pattern emerged. Comparison of wild species with cultivated varieties revealed a markedly lower level of phenolic compounds in domesticated species where about 5%–10% less of these constituents were present. Thus, germination in wheat may be related to the level of a natural chemical signal in the form of phenolic acids and their metabolites. These compounds are described in the following section.

Phenylpropanoids

The phenylpropanoids (Figure 3.4) are naturally occurring phenolic compounds that are synthesized by plants from the amino acid phenylalanine (Figure 3.5). Amino acids are considered in the later sections of this chapter in relation to proteins and the peptide bond.

The name phenylpropanoid is derived from the six-carbon, aromatic phenyl group and the three-carbon propene tail.

Phenylpropanoids are essential building blocks that are converted by plants

- Into proteins and various secondary metabolites (see Glossary)
- For use as pigments
- To form lignin for physical strength, for ultraviolet light protection and for defense against herbivores and insect attack.

Lignans

Important polyphenolic compounds, known as lignans, are found in species of wild cereals; *Aegilops*, *Triticum* and *Hordeum* (barley). Fiber-rich food items such as cereals as well as brassicas and fruit are good sources of lignans. Lignans are secondary metabolites (see Glossary). The structures of different lignans arise from various combinations of dimers of phenylpropanoids. Three common phenylpropanoid

(1)

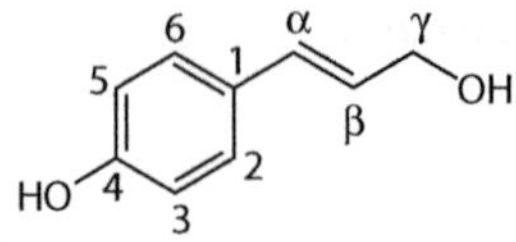

(2)

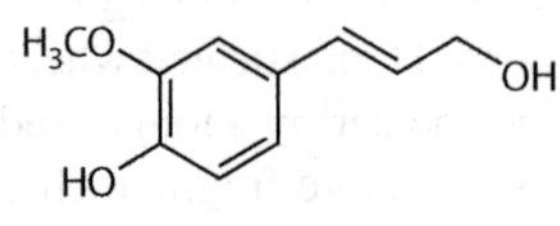

(3)

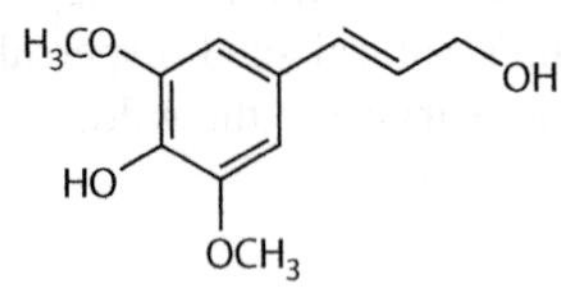

FIGURE 3.4 Three important phenylpropanoid compounds precursors of lignan and lignin: paracoumaryl alcohol (1), coniferyl alcohol (2) and sinapyl alcohol (3).

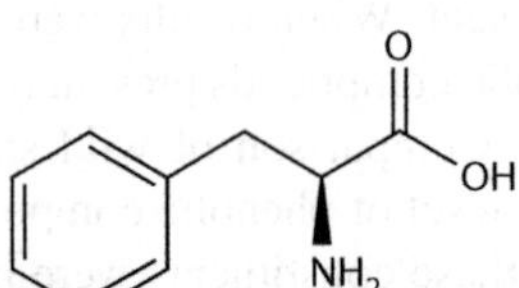

FIGURE 3.5 Structure of the amino acid, phenylalanine.

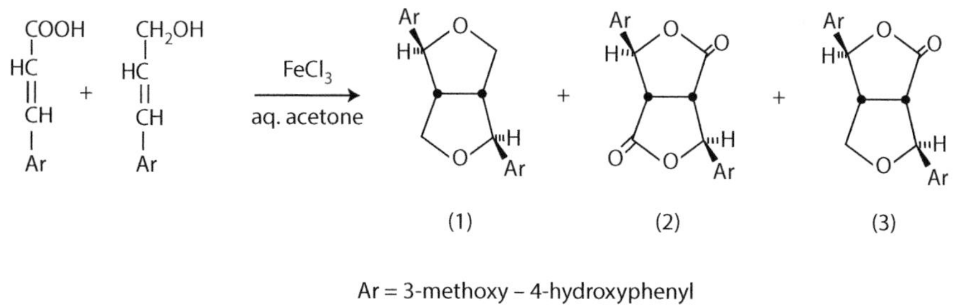

FIGURE 3.6 Coupling of two different phenylpropanoids to create lignan structures. (Taken from R. Cooper. 2015. *Journal of Traditional and Complementary Medicine*, 5:138–143.)

compounds are paracoumaryl alcohol (1), coniferyl alcohol (2) and sinapyl alcohol (3) (see Figure 3.4). They serve as natural precursors (natural building blocks) of lignans, and illustrated by the chemical reaction shown in Figure 3.6.

Although not scientifically proven in wheat species, it is believed that this is the mechanism by which two phenylpropanoids are able to form a lignan. Research has revealed that lignans behave as important germination regulators (Figure 3.6) acting as natural "rain gauges." Understanding this property was especially important for the development of modern wheat crops. Further, it may explain how the ancient species survived and germinated in arid regions.

Lignin

Phenylpropanoids are not only precursors of lignans but also of the very large polymer, known as lignin, which is commonly found in plants. Lignin is abundant in nature as it is found in the cell walls of all green plants. The formation of the polymer matrix is based upon multiple combinations (condensations) of these phenylpropanoid building blocks. Together, lignin and cellulose provide a vital structural function in plants, analogous to that of epoxy resin and glass fiber in a fiberglass boat. Lignin provides rigidity while cellulose bears the physical load. Grasses, including cereals, have lignin content less than 20% by weight and are pliable, easily bending under their own weight. However, trees are able to grow much taller and more rigidly due to additional lignin content of up to 30% by weight.

Major Cultivated Species of Wheat

Common wheat *(Triticum aestivum),* typically used in bread, is the most widely cultivated species in the world. Durum wheat (*Triticum durum*) is the second most widely cultivated wheat. Einkorn (*Triticum monococcum*) was domesticated at the same time as emmer wheat (*Triticum dicoccum*), but neither of these species is in widespread use. Spelt (*Triticum spelta*) is cultivated in limited quantities.

While each individual species has its own peculiarities, the cultivation of all cereal crops is similar. Most are annual plants; consequently, one planting yields one harvest. Wheat, rye, triticale, oats, barley and spelt are the cool-season cereals.

These are hardy plants that grow well in moderate weather and cease to grow in hot weather (~30°C). Barley and rye are the hardiest cereals, able to overwinter in subarctic regions such as parts of Siberia. Many cool-season cereals are grown in the tropics but only in cooler highlands where it may be possible to grow multiple crops in a year.

The warm-season cereals are tender and prefer hot weather. They are grown in tropical lowlands year-round and in temperate climates during the frost-free season. Rice is commonly grown in flooded fields though some strains are grown on dry land. Other warm-climate cereals, such as sorghum, are adapted to arid conditions.

Modern Wheat

The three main cash crop cereals in the world today are wheat, rice and maize. Wheat (Figure 3.7), which originated in the Levant region of the Near East and in the Ethiopian Highlands, is now cultivated worldwide. The grain has always provided an important source of vegetable protein in human food. The grain was easily cultivated, particularly on a large scale, and could be stored after harvest. This source of food enabled settlements to be established at the start of civilization as populations grew in the Babylonian and Assyrian empires known as the "Fertile Crescent."

Wheat has the ability to self-pollinate, and this attribute greatly facilitated the selection of many distinct domestic varieties. Wheat was used to make flour for baked breads, and eventually, its use spread to biscuits, cakes and in modern times to breakfast cereal, pasta and noodles. Wheat is used in the fermentation process to make beer and other alcoholic beverages, and also biofuel. Also, it provides food for

FIGURE 3.7 Modern wheat. (Taken from R. Cooper. 2015. *Journal of Traditional and Complementary Medicine*, 5:138–143.)

domestic livestock. In England, stalks of wheat were also widely put to use as thatch for roofing from the Bronze Age onward even up to the present day.

Emmer wheat (*T. dicoccum*) is also known as faro in Italy. The main use of emmer is as a human food although it is also used for animal feed. Ethnographic evidence from Turkey suggests that emmer makes good bread, and this is supported by evidence of its widespread consumption as bread in ancient Egypt. Today, emmer bread (*pane de faro*) is available in Switzerland and in Italy. Emmer is higher in fiber content than common wheat and is used for making pasta. Emmer has also been used in beer production, an example being the Riedenburger ecobrewery in Bavaria, Germany which currently makes "Emmerbier." As with most varieties of wheat, however, emmer is probably unsuitable for sufferers from wheat allergies or celiac disease (see the section on "Rediscovering Traditional Grains of the Americas: Chia and Quinoa").

Wheat Genetics

Some wheat species have two sets of chromosomes and are known as diploid, whereas many are tetraploid with four sets of chromosomes, or hexaploid with six. Einkorn wheat (*T. monococcum*), for example, is diploid. Most tetraploid wheats (emmer and durum wheat) are derived from wild emmer, *Triticum dicoccoides*. Wild emmer is itself the result of hybridization of two diploid wild grasses, *Triticum urartu* and a wild goat grass such as *Aegilops searsii* or *Aegilops speltoides*. The exact identity of the grass has never been established, but among surviving wild grasses the closest living relative is *A. speltoides*. The hybridization that formed wild emmer occurred in the wild long before domestication, and was driven by natural selection. Hexaploid wheat evolved in farmers' fields. Domesticated emmer, or durum wheat, when hybridized with another form of wild diploid grass (*Aegilops tauschii*) produces hexaploid wheats, known as spelt or bread wheat. These have three sets of paired chromosomes, three times as many as in diploid wheat.

There is some evidence that gluten from einkorn contains an isomeric form of the protein, gliadin, which may not affect sufferers of celiac disease as seriously as the gliadin found in the gluten of modern strains of wheat. Einkorn grain could make bread, which is more palatable to sufferers who are obliged to follow a gluten-free diet. That einkorn is distinct from other wheat varieties in this way is no doubt due to genetic differences—einkorn has only 14 chromosomes whereas emmer possesses 28 chromosomes, and modern wheat 42 chromosomes (see also the section on "Rediscovering Traditional Grains of the Americas: Chia and Quinoa" for more on gluten sensitivity and celiac disease).

Nutritional Importance of Wheat and Grinding

Wheat contains protein, fat, carbohydrate, dietary fiber (in the form of starch) and iron. Wheat starch is considered an important commercial by-product of wheat and second only in economic value to the wheat gluten.

When consumed in their natural whole-grain form, cereal grains are a rich source of vitamins, minerals, carbohydrates, fats, oils and protein. However, when refined by the removal of the fibrous parts of bran and germ, the remaining endosperm is mostly carbohydrate and lacks the majority of these nutrients.

The whole grain can be milled to leave just the endosperm for white flour. The by-products of this process are bran and germ. The whole grain is a concentrated source of vitamins, minerals and protein, while the refined grain is mostly starch. The four wild species of wheat, along with the domesticated varieties einkorn, emmer and spelt, have toughened husks (glumes) that tightly enclose the grains. In domesticated wheat varieties, the rachis, which is brittle, breaks easily on threshing. (In plants, the rachis is the main axis of a compound structure such as the main stem of a compound leaf.) As a result, when threshed, the ear of wheat breaks up into spikelets. In order to obtain the grain, further processing, such as milling or pounding, is needed to remove the husks. In contrast, in free-threshing forms such as durum wheat and common wheat, while the husks are fragile, the rachis is tough. Husked varieties of wheat are often stored as spikelets, because the toughened husks give good protection against pests of stored grain.

The protein content of bread wheat ranges from 10% in some soft wheat varieties with high starch content to 15% in hard wheat varieties. However, the value of the wheat protein is influenced by its gluten content, which will determine its applicability to a particular dish. The gluten present in bread wheat is strong and elastic and enables the dough made from it to trap carbon dioxide during leavening. However, elastic gluten also interferes with the rolling of pasta into thin sheets so gluten protein from durum wheat is used for pasta instead as this is strong, but not elastic.

Importance of Proteins

Whole grains from such staple food as cereals are a source of protein. Proteins are found in *Triticum*, wheat, oats, rye, millet, maize (corn) and rice.

Proteins are essential nutrients for the human body and can also serve as an energy source. Protein can be found in all cells of the body. It is the major structural component of cells, especially those of muscle.

Proteins are polymer chains made from amino acids linked together by peptide bonds. During human digestion, proteins are broken down into smaller polypeptide chains. This process of release of essential amino acids is crucial since they cannot be biosynthesized by the body.

Enzymes are proteins too, but they are of a particular type with a highly specialized function. An enzyme is a metabolic catalyst, which will only act with a given substrate, which has to fit exactly into an active chemical site on the enzyme. This is rather like a skillfully cut key fitting precisely into a well-made lock such that the lock can be operated at will (Figure 3.8). If the substrate fits exactly into the active site in each one of three dimensions in space, metabolism will proceed, otherwise it cannot.

The substrate is held in place on the active site without chemical binding by weak close-range attraction, due to a combination of hydrogen bonding and van der Waals forces. Once the substrate has been captured in this way, the substrate can react more readily with other compounds present. This is because the binding of the substrate to the enzyme causes redistribution of the electrons in the chemical bonds of the

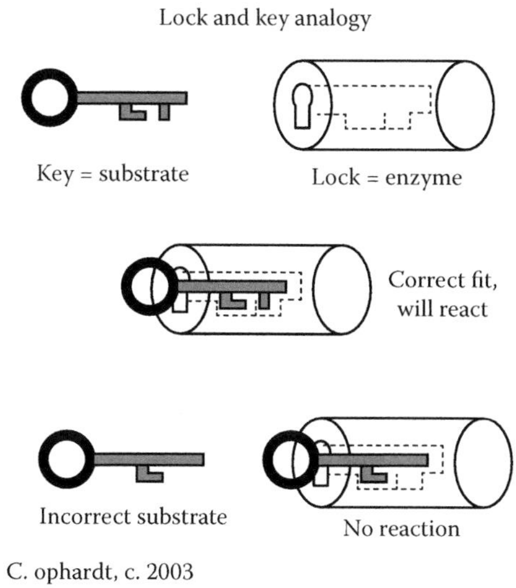

FIGURE 3.8 **(See color insert.)** Schematic illustration of an active site in an enzyme—the lock and key model. (From http://www.elmhurst.edu/~chm/vchembook/571lockkey.html)

substrate. This event triggers reactions in the substrate, leading to the formation of products, which are released, thus allowing the active site of the enzyme to engage in another reaction cycle.

Peptide Bond

A peptide bond (Figure 3.9) is a covalent chemical bond formed between two molecules when the carboxyl group of one molecule reacts with the amino group of another molecule, causing the release of a molecule of water. This reaction occurs because amino acids are amphoteric, that is, they possess both alkaline properties (due to an amine functional group) and acidic properties (due to a carboxylic functional group) at one and the same time (see also the sections on "Tobacco: A Profound Impact on the World," for more information on amines, and "Europe Solves a Headache," for more material on carboxylic acids).

The following reaction between two molecules to form the most simple of amino acid is illustrative:

$$2NH_2COOH = NH_2CONHCOOH + H_2O$$

The resulting C(O)NH group is called a peptide bond or link. Peptide bonds can be formed repeatedly between the carboxyl group and the amino group of neighboring amino acids. Proteins are polymers formed from chains of peptide bonds.

The Condensation Reaction

The process of forming a peptide bond is an example of a condensation reaction. In a condensation reaction, the molecules of two compounds combine together with the

$$2NH_2COOH = NH_2CONHCOOH + H_2O$$

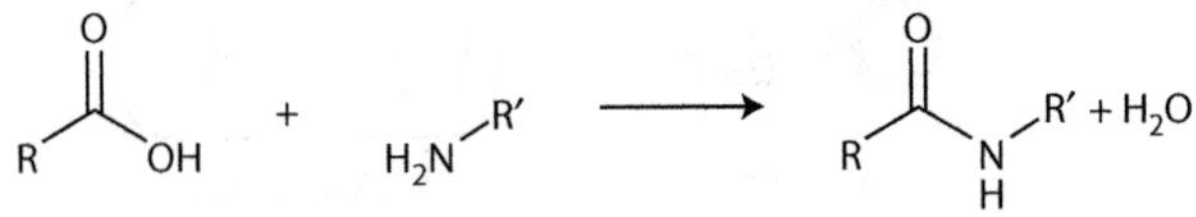

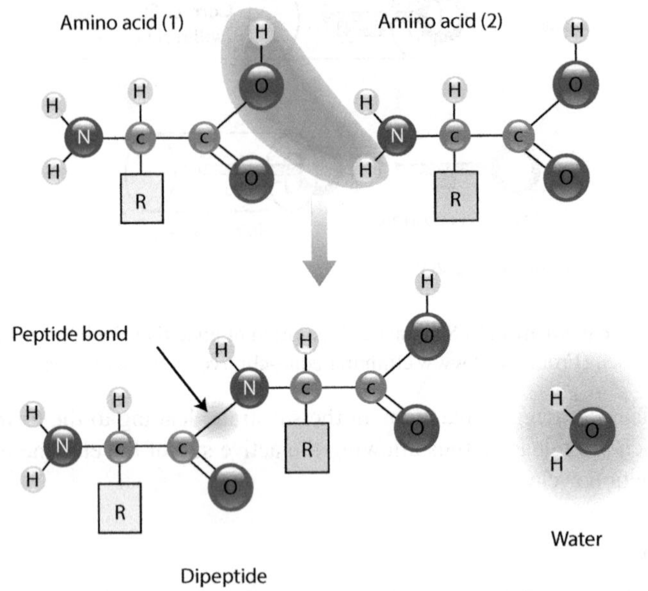

FIGURE 3.9 **(See color insert.)** Formation of the peptide bond and formation of a dipeptide.

elimination of a molecule of water. There are many examples of condensation reactions in organic chemistry beyond the formation of polymers such as proteins. For instance, ethanoic or acetic acid forms acetic anhydride and water by the condensation reaction

$$2CH_3COOH = (CH_3CO)_2O + H_2O$$

This reaction in reverse would be an example of hydrolysis.

Proteins and Amino Acids

As we have seen earlier in this section, proteins are polymers made up from building blocks known as amino acids which are linked together by peptide bonds. Over 500 different amino acid structures have been identified in nature. However, the human genetic code relates to only 20 of these. In other words, every protein in the human body consists of some combination of only these 20 amino acids. They can be sorted into two groups: essential and nonessential. Essential amino acids must be obtained

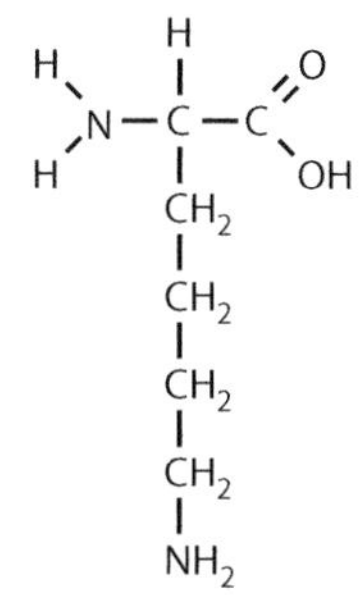

FIGURE 3.10 The molecular structure of lysine—an example of an essential amino acid.

from the diet, whereas nonessential amino acids are those which the human body is capable of synthesizing.

The names of essential amino acids are histidine, isoleucine, leucine, lysine (Figure 3.10), methionine (Figure 3.24), phenylalanine, threonine, tryptophan and valine.

The nonessential amino acids are as follows: alanine, arginine, asparagine, aspartate, cysteine (Figure 3.24), glutamic acid, glutamine, glycine, proline, serine (Figure 3.11) and tyrosine (Figure 3.9).

Amino acids found in the body cannot be stored in the same manner as fat or starch and must be obtained regularly from foods, which contain proteins. Protein intake is therefore vital. The amino acids are obtained from the breakdown of proteins, which are present in the tissues of either animals or plants. The legume family is an example of a good source of dietary protein from plants.

The Molecular Structure of Proteins and NMR

A common goal of research is to obtain a high-resolution three-dimensional structure of a protein. This can be achieved by the use of x-ray crystallography or another important technique, nuclear magnetic resonance (NMR) spectroscopy (see the section on "Saving the Pacific Yew" for an explanation of the principles behind NMR spectroscopy).

The active site of an enzyme, which is a protein, can be better understood using NMR. For example, let us consider the way modern drugs interact with our body. Many effective drugs are enzyme inhibitors (see Glossary). The mechanism of action of these drugs can be described using the model of a lock and key. If the details of

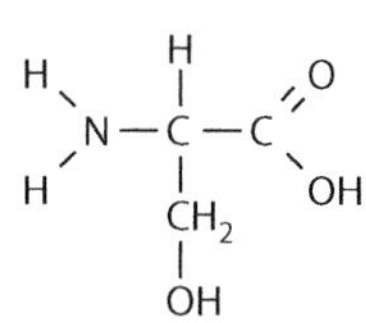

FIGURE 3.11 The molecular structure of serine—an example of a nonessential amino acid.

the structure of the protein are examined using NMR and are compared before and after interaction with the trial drug, then the nature and site of biochemical activity within the molecule can be deduced.

NMR spectroscopy has been used also to help to elucidate the structures and isomers of other complex molecules, such as nucleic acids and carbohydrates.

Fascinating Influence of Chirality on Life Systems on Earth

In the section on "Saving the Pacific Yew," attention was given to the concept of chirality in molecules, as this is of fundamental importance in natural products chemistry.

It is fascinating to note that only left-handed amino acids are found in the life systems of plants and animals. As a consequence, only left-handed proteins (and enzymes) are found in the natural world. Outside the natural world, nucleotides are very uncommon, but when they are found they exist in both left-handed and right-handed molecular structures in approximately a ratio of 1:1.

Life on Earth could have emerged from either left-handed or right-handed amino acids but not both, as metabolic processes and pathways are structure specific and are not possible with a mix of left-handed and right-handed proteins and enzymes.

Is this chance or design? No one knows for certain! It is a conundrum, which has intrigued scientists for many years and has stimulated much debate. Although unproven theories abound, some perhaps rather close to speculation, sheer chance seems the most likely explanation. A suggestion for further reading is given at the end of the chapter.

The Shikimic Acid Pathway

The shikimic pathway is found in green plants, bacteria and fungi but not in mammals. The basic building block is shikimic acid (Figure 3.12).

The shikimic pathway is shown in Figure 3.13 and involves seven enzymes acting in sequential steps to catalyze the biosynthesis of three aromatic amino acids; phenylalanine, tyrosine and tryptophan.

As well as the amino acid functional group, phenylalanine also contains a phenyl functional group; tyrosine also has a phenol functional group, whereas tryptophan differs through the presence of the indole functional group.

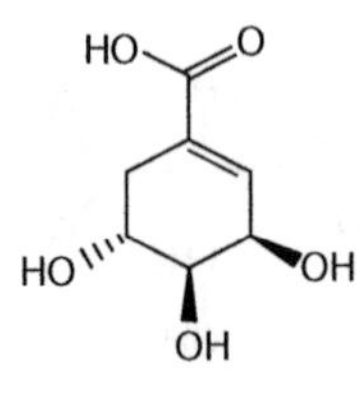

FIGURE 3.12 Structure of shikimic acid.

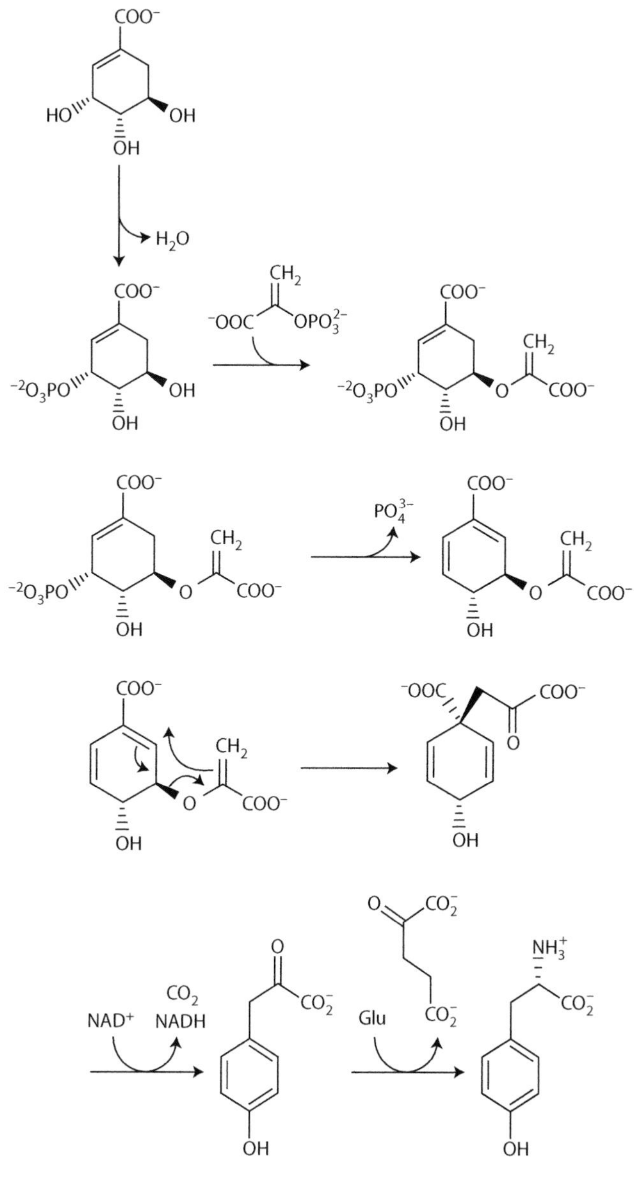

FIGURE 3.13 The shikimate pathway to generate amino acids. (Taken from R. Cooper and G. Nicola. 2014. *Natural Products Chemistry: Sources, Separations and Structures.* CRC Press, Taylor & Francis Group.)

In this manner, by the last step, the generation of an aromatic amino acid (in this case tyrosine) has been achieved. Thus, entry to the shikimic pathway generates a phenylpropanoid which, in turn, can be converted to amino acids, peptides and proteins. The structures of these key essential aromatic amino acids are shown in Figure 3.14.

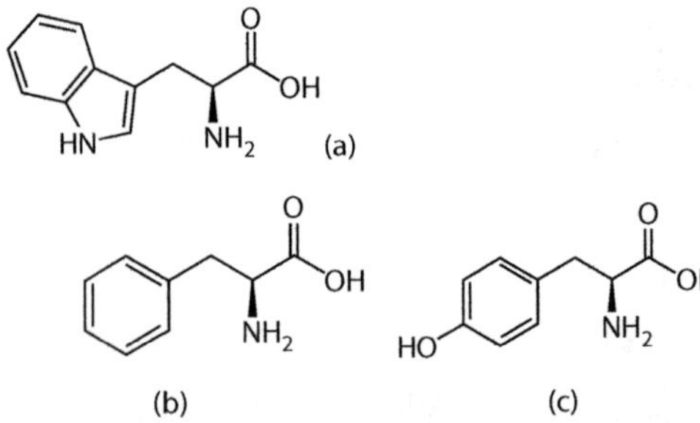

FIGURE 3.14 Examples of aromatic amino acids: tryptophan (a), phenylalanine (b) and tyrosine (c).

Eugenol and Rosmarinic Acid

Two important natural oils namely, eugenol and rosmarinic acid (Figure 3.15), are also phenylpropanoids. Both eugenol and rosmarinic acid possess the characteristic "phenolic head" group, which is attached to a 3-carbon tail.

Eugenol has been used as a dental analgesic and antiseptic. This oil is found in concentrations of 80%–90% in clove oil.

Society's Challenges in Enhancing Agricultural Production: Crop Yield and "Green Issues" Including Genetic Modification of Crops

In crop production today, an urgent need exists to improve crop yield, almost in a desperate attempt to feed a hungry world given the ever-expanding human population.

Over time, in more developed parts of the world, nitrate and phosphate fertilizers have been applied liberally to the soil to stimulate quicker and thicker growth in cereal crops. To some extent this has been successful, but at a cost to the environment. Over reliance on chemical fertilizers produced in bulk by the inorganic chemical industry has led to diminished soil conditions as natural processes, which involve microorganisms and improve fertility, are interrupted. The soil also becomes compacted under the weight of modern, heavy farm machinery. The runoff of rain water from fields pollutes water courses and lakes. Excess dissolved fertilizer promotes the growth of algal blooms, which reduce the light available to other green plants and drastically alters the balance of dissolved gases in the water, such as oxygen and carbon dioxide, thereby choking fish and other invertebrates. These changes affect the food chain of other wild creatures (such as mammals, otters and birds, herons

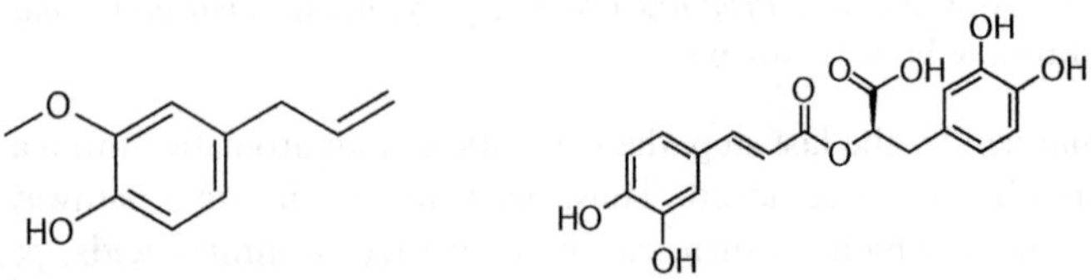

FIGURE 3.15 Molecular structures of eugenol and rosmarinic acid.

and water rails) which depend on water life as a major food resource. Water quality for human use is also adversely affected.

Pesticides (insecticides and herbicides) continue to be applied in large quantities to maintain a monoculture in the field by reducing or removing competition for crop tissues and light and space. These measures undoubtedly harm wildlife and may have an adverse impact on human health. Mammals, including human beings, are at the top of the food chain and consume cereals and other animals, which have grown by eating grain. Some pesticides are made from chemicals, which do not degrade naturally in the soil or general environment. There is a risk that minute quantities of these stubbornly resistant chemicals enter the food chain and can build up in human tissue over time with consequent and subtle effects on human health, which are not as yet fully understood.

Genetically modified crops have been created by man from natural strains of wheat. Genes have been artificially inserted into the genome of the cereal plant in order to selectively improve qualities valuable to man in the form of crop yield, good and consistent size in grain, flavor, resistance to drought and disease; or in the length of the stalk, which, if shorter, is more harvestable as it is less likely to blow over in windy, wet weather and involves less waste product in the form of straw. However, the long-term effect on the environment is as yet unknown. Despite great improvements in the efficiency of modern farm machinery, it is clearly impossible to collect every last scrap of grain from a field. Birds and the wind are bound to carry some of this grain away to the field boundary or beyond with unknown consequences for the environment. Unintended cross-fertilization of natural plants with genetically modified crops may well produce "super" weeds, which are resistant to disease and the application of herbicides, thus creating the need for even more research and development. The energy and space requirements of genetically modified crops too are generally the same as those for naturally bred species of cereals.

Glyphosate is commercially sold worldwide—one well-known brand being Roundup®. Glyphosate is a nonselective herbicide, which means, when applied it is unspecific in its action as it has the ability to kill all plants. It works because glyphosate interferes with the shikimic pathway. Glyphosate has been used to suppress weeds in conjunction with crops, which have been genetically modified to withstand it. This approach has economic value in that the practice of tilling the soil is very time consuming and uses up energy and also reduces soil compaction rendering the soil more liable to erosion by wind and water. Since the shikimic pathway is not a part of mammalian metabolism it has been suggested that humans are immune to the effects of glyphosate. However, this may be an oversimplification in that bacteria (which are single-celled plants) are present in vast numbers in the human intestinal tract. They are essential for good health in processing certain food items and in playing a part in the vital immune system. If glyphosate were to be admitted to the intestinal system, it could suppress the beneficial behavior of these valuable bacteria as suggested in a paper by Samsel and Seneff (2013).

Food from genetically modified crops is widely consumed in the United States and Canada either directly or in processed material as a factor in soy protein, oils and flavorings. Here, food labeling does not need to draw attention to the use of genetically modified crops. In other parts of the world, notably in European countries, Australia and Japan, a degree of circumspection exists, where labeling is more explicit and in some instances food from genetically modified crops is prohibited from sale because it is not as yet, considered proven to be safe.

Unless more land is brought into cultivation, which in itself has an impact on the habitat of wildlife and biodiversity, it is difficult to envisage how crop yields can continue to rise inexorably. Careful management and application of all of these techniques, coupled with research into more effective practices and materials, which can be demonstrated to be entirely harmless to man, continue to be required.

QUESTIONS

1. What is the difference between lignin and lignan? Where is lignin found in plants and why is it important to the plant's growth?
2. Give an account of how lignans act in inhibiting the germination of wheat.
3. In modern farming, benefits should outweigh risks in order to justify the mass production of food items in a particular way. Give a balanced account of the pros and cons involved in the use of genetically modified crops.
4. Describe the chemical properties you would expect of phenylalanine given the functional groups present in the molecule.
5. Phenylalanine is a significant building block in the shikimic acid pathway through which secondary metabolites are formed in plants. Comment upon the differences between primary and secondary metabolites. Explain the particular interests which humankind has in both.
6. Figures 3.10 and 3.24 show lysine and methionine respectively, which are examples of essential amino acids. Look up and draw the structures of the other essential amino acids; histidine, isoleucine, leucine, phenylalanine, threonine, tryptophan and valine.
7. Similarly, three nonessential amino acids are shown in the chapter: serine (Figure 3.11), tyrosine (Figure 3.9) and cysteine (Figure 3.24). Look up and draw structures for the other nonessential amino acids: alanine, arginine, asparagine, aspartate, glutamic acid, glutamine, glycine and proline.

REFERENCES

R. Cooper. 2015. Re-discovering ancient wheat varieties as functional foods. *Journal of Traditional and Complementary Medicine*, 5:138–143.

R. Cooper, D. Lavie, Y. Gutterman, M. Evanari. 1994. The distribution of rare phenolic type compounds in wild and cultivated wheats. *Journal of Arid Environments* 27:331–336.

S. Lev-Yadun, A. Gopher, S. Abbo. 2000. The cradle of agriculture. *Science* 288:1602–1603.

D. Pizzuti, A. Buda, A. d'Odorico, R. d'Incà, S. Chiarelli, A. Curioni, D. Martines. 2006. Lack of intestinal mucosal toxicity of *Triticum monococcum (Einkorn)* in celiac disease patients. *Scandinavian Journal of Gastroenterology* 41:1305–1311.

A. Samsel and S. Seneff. 2013. Glyphosate suppression of enzymes and amino acid biosynthesis. *Entropy* 15(4):1416–1463.

SUGGESTED FURTHER READING

J. Diamond. 1997. *Guns, Germs and Steel: A Short History of Everybody for the Last 13,000 Years.* Random House.

N. Eviatar, A. B. Korol, A. Beiles, T. Fahima. 2002. *Evolution of Wild Emmer and Wheat Improvement: Population Genetics, Genetic Resources, and Genome.* Springer.

S. T. Grundas 2003. Wheat: The crop. In *Encyclopedia of Food Sciences and Nutrition.* Page 6130. Elsevier Science Ltd.

J. F. Hancock. 2004. *Plant Evolution and the Origin of Crop Species.* CABI Publishing.

M. Hopf and D. Zohary. 2000. *Domestication of Plants in the Old World: The Origin and Spread of Cultivated Plants in West Asia, Europe, and the Nile Valley*, 3rd edition. Page 38. Oxford University Press.

A. L. Kolata. 2009. Quinoa: Production, Consumption and Social Value in Historical Context. Department of Anthropology. University of Chicago.

A. Kessel and N. Ben-Tal. 2012. *Introduction to Proteins: Structure, Function and Motion.* Taylor & Francis.

L. A. Morrison. 1994. Re-Evaluation of Systematic Relationships in *Triticum L* and *Aegilops L* Based on Comparative Morphological and Anatomical Investigations of Dispersal Mechanisms. Thesis (PhD). Oregon State University.

ASIAN STAPLE: RICE

Abstract: Rice feeds a greater proportion of the world's population than any other cereal crop. Cereal grain is rich in carbohydrates, which are plentiful in nature. Carbohydrates occur widely in both the plant and animal kingdoms where they are present in supportive tissues and are a source of energy.

Natural products chemistry

- Carbohydrates and oxygen in the organic ring
- Monosaccharides (sugars, glucose and fructose) as natural building blocks
- Disaccharide sugars, sucrose and lactose
- Oligosaccharides
- Polysaccharides, cellulose and starch.

Curriculum content

- Aldehydes (alkanals)
- Ketones (alkanones)
- Cyclic compounds.

Carbohydrates and Saccharides: Oxygen in the Organic Ring

In organic chemistry, this very large class of substances is usually referred to in general terms as the carbohydrates, whereas in biochemistry, they are typically known as the saccharides. The origin of the term, saccharide, is the Greek name for sugar, "sacchar." Incidentally, the name saccharin has been given to a man-made, artificial sweetener which is neither a sugar nor a carbohydrate, but has the functional group of an amine within a five-member carbon ring, which is integral to a benzene ring.

Carbohydrates (see Table 3.1) are the most abundant class of organic compounds found in living organisms. The generic name, carbohydrate (*carbon hydrates*), arises from the observation that the molecular formula of this class of compounds is $C_n(H_2O)_n$, where n is typically a large number. Carbohydrates are produced by photosynthesis: the condensation of carbon dioxide requiring energy, in the form of

TABLE 3.1
The General Classification of Carbohydrates

Carbohydrates		
Sugars	**Nonsugars**	
Monosaccharides	Oligosaccharides	Polysaccharides
Pentoses Hexoses	Disaccharides (Tri and tetra)	Homo Hetero
(Glucose, fructose)	(Maltose, sucrose)	(Cellulose, starch)

light, and the green pigment, chlorophyll (see the section on "Our World of Green Plants: Human Survival" in Chapter 7 for more details).

$$nCO_2 + nH_2O + energy \rightarrow C_nH_{2n}O_n + nO_2$$

Monosaccharides, whether linear or cyclic, can be linked together in many different ways. Thus, a huge variety of individual carbohydrates and their isomers exist as disaccharides, oligosaccharides and polysaccharides.

Carbohydrates may be subdivided broadly into two groups; sugars and nonsugars. Sugars, given the suffix, *ose*, are monosaccharides or disaccharides. Sugars are colorless, crystalline solids that are soluble in water and usually have a sweet taste. In contrast, nonsugars are white, amorphous solids insoluble in water. Nonsugars can be broken down by chemical hydrolysis into their constituent sugars, which is why nonsugars are known as polysaccharides. If only one type of sugar is produced by hydrolysis, the nonsugar is known as a homopolysaccharide, whereas if more than one different sugar is released it is described as a heteropolysaccharide.

Monosaccharides

The simplest carbohydrates are called monosaccharides. They may be linear or may have either a six-membered or a five-membered ring structure. Representative examples are glucose and furanose, respectively (in Figure 3.16).

Even glucose, a monosaccharide, can exist in several isomeric forms (Figure 3.17), owing to the fact that the hydroxyl group can be in different configurations on each carbon atom in the ring.

Disaccharides

When two monosaccharide molecules are joined together through a glycoside bond, they can form a disaccharide molecule. When more monosaccharide molecules link together, then a polymer called a polysaccharide is formed. Two disaccharides occurring widely in nature with the molecular formula, $C_{12}H_{22}O_{11}$, are sucrose (Figure 3.18), the sugar of plants, and lactose (Figure 3.18), the sugar of animals. The molecular structure of sucrose indicates how glucose is linked to fructose through a glycoside bond. Given that the systematic names of saccharides can be very long, arbitrary names remain in use for well-known substances.

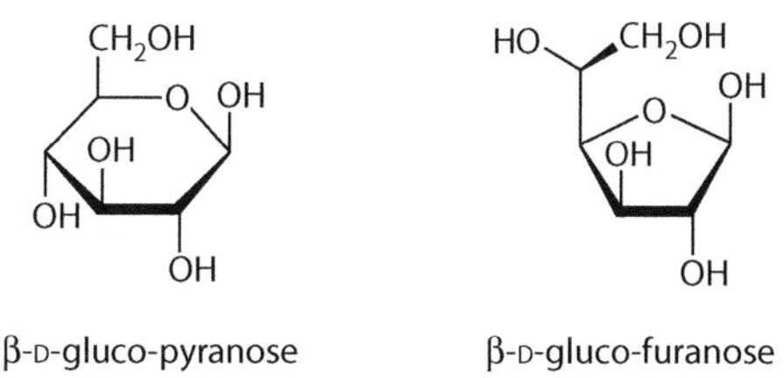

FIGURE 3.16 Glucose and furanose.

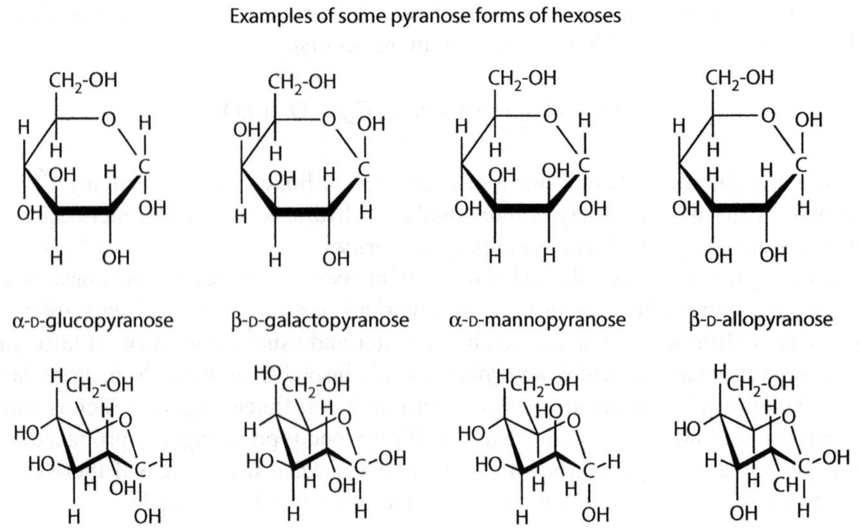

FIGURE 3.17 Examples of isomers of glucose. (Taken from R. Cooper and G. Nicola. 2014. *Natural Products Chemistry: Sources, Separations and Structures.* CRC Press, Taylor & Francis Group.)

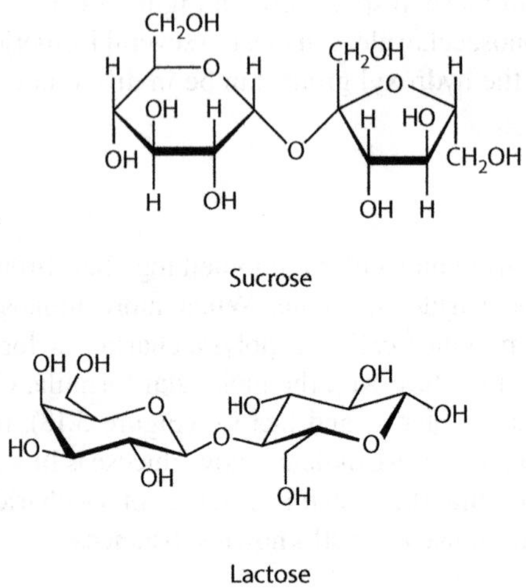

FIGURE 3.18 Structural formula of sucrose and lactose.

The Glycoside Link

The glycoside link is the name given to the ether-like bond involving the α-carbon atom 1 in the six-member ring and the β-carbon atom 2 in the five-member ring, for example in the sucrose molecule.

Sucrose, commonly recognized as table sugar, is abundant as it is found universally in plants and in honey, which is derived by bees from plant material. Sucrose crystallizes readily from aqueous solution. Commercially, sucrose is obtained from the stem of the sugar cane grown in tropical lands, and from the root of the sugar beet, cultivated in temperate parts of the world, notably northern Europe.

Lactose is the sugar found in the animal kingdom, and is present in small concentrations in blood plasma and in the milk of mammals. It may be obtained commercially by evaporating whey, the aqueous liquor remaining as a by-product in the making of cheese.

Oligosaccharides

Oligosaccharides are intermediate in size. Oligosaccharides are composed of several monosaccharides joined through glycoside links. As with other polysaccharides, the glycoside links can be hydrolyzed in the presence of enzymes or acid to yield the constituent monosaccharide units. A carbohydrate consisting of 2–10 monosaccharide units with a defined structure is regarded as an oligosaccharide—anything larger would be referred to as a polysaccharide.

Polysaccharides

Very large polymeric molecules, made up from a high number and/or a great variety of monosaccharides, are known as polysaccharides. It is easy to appreciate that polysaccharides can attain very high molecular weights of over 100,000 Da (see Glossary). Polysaccharides are a major source of metabolic energy, both for plants and for those animals, which depend on plants for food. Polysaccharides are a component of the energy transport compound, adenosine triphosphate (ATP) (see the Glossary and the section on the "Chinese Cordyceps: Winter Worm, Summer Grass"), and are found on the recognition sites of cell surfaces (see proteins and enzymes in the section on "Foods of the Fertile Crescent: Ancient Wheat").

Cellulose is a good example of a homopolysaccharide and is represented by the formula $(C_6H_{10}O_5)_n$ where n ranges from 500 to 5,000 depending on the natural source of the polymer. Cellulose is found in wood (50%), jute (70%), hemp and flax (80%), and in the seed hairs of the cotton plant (almost 100%). Starch is also a homopolysaccharide. While cellulose is completely insoluble in water, starch is only partially soluble in water.

Starch is the substance in which plants store their reserves of carbohydrate and is typically found in bulbs, tubers and seeds. The main commercial sources of starch are rice, wheat, maize and potatoes. The homopolysaccharide, starch, may be hydrolyzed under suitable conditions to glucose.

Another homopolysaccharide is inulin—also represented by the generic formula $(C_6H_{10}O_5)_n$. Inulin yields fructose on hydrolysis and is obtained from the tuber of the dahlia. The dahlia comes to mind as an ornamental plant originating in Mexico, where it is regarded as a national emblem, but an important constituent of dahlia tubers is digestible carbohydrate together with polysaccharide fiber and some protein. The Aztecs consumed dahlia tubers as a source of food carbohydrate, although the polysaccharide, inulin, is indigestible by bacteria resident in the human alimentary canal and is thus, excreted as fiber.

Yet another important group of carbohydrates is the heteropolysaccharides. Galactomannans are an example of a heteropolysaccharide involving the sugars, galactose and mannose, within the polymer—hence the name of the class. Galactomannans may be added to processed food products such as ice cream to increase viscosity and hence improve texture (for more on cellulose and galactomannan see the section on "Global Aloe" in Chapter 6).

Chemical Properties of Carbohydrates: Monosaccharides

This book illustrates the chemical properties of carbohydrates through the relatively simple and common monosaccharides, glucose and fructose. The molecular formula of both is $C_6H_{12}O_6$. Although both sugars are good sources of energy, excess of glucose can be fatal to diabetic patients and excess of fructose in the human body can lead to health problems related to liver disease and to insulin resistance.

While the name carbohydrate might suggest that these compounds are "hydrates of carbon" and might behave as such, in reality they are more accurately described as polyhydroxy aldehydes and polyhydroxy ketones.

A linear molecule of glucose has the form

$$CH_2(OH)\cdot CH(OH)\cdot CH(OH)\cdot CH(OH)\cdot CH(OH)\cdot CHO$$

although the aldehyde group could, of course, be present on any one of the carbon atoms along the chain. It is not surprising that this form of glucose displays the organic chemistry of the alkanals (or aldehydes) and alcohols. Given the presence of the aldehyde functional group, glucose is a reducing agent and is known as an aldohexose. It is a most important aldose.

In linear form, fructose has the structure

$$CH_2(OH)\cdot CH(OH)\cdot CH(OH)\cdot CH(OH)\cdot CO\cdot CH_2OH$$

While the ketone group could, of course, be present on any one of the carbon atoms along the chain, it is interesting to note that in naturally occurring fructose, the carbonyl functional group is located on the second carbon atom. It is not surprising, therefore, that this form of fructose displays the organic chemistry of the alkones (or ketones) and the alcohols. Given the presence of the ketone functional group, fructose can react with hydrogen cyanide and amines. Fructose, therefore, is described as a ketohexose and is quite abundant, occurring in free form in many fruit juices and in honey.

It must be emphasized that the ring molecules of both glucose and fructose do not have either of these functional groups since the ring is closed by an ether linkage of carbon to oxygen to carbon. Also, it is worth pointing out that a number of isomers of each substance exist in the cyclic form of the molecule.

Under appropriate conditions, the hydroxyl moiety of glucose or fructose can undergo reactions typical of a hydroxyl functional group:

- Elimination of water to form the disaccharide, sucrose
- Esterification with, for example, acetic anhydride to form an acetate and ultimately a penta-acetate

- Reaction with either an aldehyde or a ketone involving neighboring hydroxyl groups to form a five-member cyclic acetal or a five-member cyclic ketal.

The Value of Monosaccharides, Disaccharides and Polysaccharides to Living Organisms

Monosaccharides are the major source of fuel for metabolism, being used both as an energy source (glucose being the most important in nature) and in biosynthesis. When monosaccharides are not immediately needed by many cells they often are converted into polysaccharides. In many animals, including humans, the polysaccharide in question is glycogen, which is retained in liver and muscle cells and can be readily converted back to glucose when energy is needed. In plants, starch is used for storage of surplus monosaccharides and for conversion back into energy. The most abundant polysaccharide is cellulose, which is the structural component of the cell wall of plants and many forms of algae. Fructose, or fruit sugar, is found in many plants. The sugar, ribose, is a component of RNA (see Glossary). Deoxyribose is a component of DNA (see Glossary).

Human Nutrition

Grain is a rich source of carbohydrates. Carbohydrates are a plentiful source of energy in living organisms although carbohydrates are not an essential component of human diet as human beings are able to obtain most of their energy requirement from protein and fats.

Organisms cannot metabolize every type of carbohydrate to yield energy, but glucose is an accessible source of energy. While many organisms can easily break down starch into glucose, most organisms cannot metabolize cellulose. However, cellulose can be metabolized by some bacteria. Ruminants exploit microorganisms in their gut and termites farm microorganisms to process cellulose. Even though these complex polysaccharides are not very digestible, they do represent an important dietary element for human beings, known simply as dietary fiber. The "rough" nature of dietary fiber aids digestion.

In food science, the term carbohydrate relates to food that is particularly rich in the complex carbohydrate, starch, such as cereals, bread and pasta, or to simple carbohydrates, such as the sugars present in candies, jam and desserts. High levels of carbohydrate are often associated with highly processed foods or refined foods made from plants; sweets, cookies, candy, table sugar, honey, soft drinks, jam, bread, pasta and breakfast cereals. Lower amounts of carbohydrate are usually associated with unrefined foods, including brown rice and unrefined fruit.

Commercial Uses of Carbohydrates

Brewing of Beer

The starch from grain of the cereal, barley, is used in the brewing industry. The process involves a number of steps.

Initially, the grain is kept in moist, warm conditions. Starch is converted to the sugar or disaccharide, maltose, having the molecular formula $C_{12}H_{22}O_{11}$, by the enzyme, diastase, which is present in barley.

$$2(C_6H_{10}O_5)_n + nH_2O = (C_{12}H_{22}O_{11})_n$$

Water is applied to dissolve the maltose and yeast is added to the liquor. Yeast contains two enzymes, maltase and zymase, which bring about fermentation in the final two steps to make beer.

$$C_{12}H_{22}O_{11} + H_2O = 2C_6H_{12}O_6$$

Here, maltose is converted to the monosaccharide, glucose, by maltase.

$$C_6H_{12}O_6 = 2C_2H_5OH + 2CO_2$$

Finally, ethanol is generated from glucose through the action of zymase. For more on the process of fermentation, see the section on "Chinese Cordyceps: Winter Worm, Summer Grass" also presented in this chapter.

Adhesives and Stiffening Agents in Fabrics

Starch is also widely employed in the preparation of adhesives and in stiffening agents for textiles and fabrics.

Bioplastic

As the cost of oil rises and the effects of global warming intensify, governments and industries are forced to turn attention to viable, natural alternatives. Bioplastics are made partly or wholly from materials derived from biological sources such as sugar cane, potato starch or cellulose from trees and straw.

Bioplastics are also often biodegradable and are, therefore, a much more sustainable product and enrich the soil on decomposition. Plant-based bioplastics can also be recycled. There is the additional benefit that biomass feedstock absorbs carbon dioxide as it grows thereby lowering the carbon footprint of the final product. Products and packaging made from bioplastics also have direct appeal to consumers.

Starch is used to produce various bioplastics, synthetic polymers that are biodegradable. An example is polylactic acid based on glucose from starch.

Cellophane is a thin, transparent sheet made from regenerated cellulose. Cellophane has low permeability to air, oils, bacteria and water, which makes it very useful for food packaging. Cellophane is totally biodegradable and is effectively a polymer of glucose similar to cellulose. Cellulose film has been manufactured continuously since the mid-1930s and is still used today. In the United Kingdom and in many other countries, "cellophane" is a registered trademark and is the property of Innovia Limited. However, in the United States the term is often used informally and more generally to refer to a wide variety of plastic wrapping films, which are not composed of cellulose.

Explosives

When cellulose is immersed in a mixture of nitric and sulfuric acids, the hydroxyl functional groups of cellulose are replaced by the nitrate group with the elimination of water. As the reaction involves an alcohol and an acid, it is an example of esterification represented simply by the equation; cellulose + acid = cellulose ester + water.

When up to six of the hydroxyl functional groups are replaced on each glucose unit, cellulose nitrate may be explosively ignited in air and is known commonly as gun cotton. The explosion produces a low pressure shock wave. Cellulose nitrate (or nitro cellulose, as it is commonly known) is a constituent of cordite, which was the material used to propel projectiles, bullets or shells, until it was superseded by other products after the end of World War II.

QUESTIONS

1. Explain what is meant by the term, carbohydrate, and describe how this broad concept may be further classified into subgroups of compounds. Where do the following substances appear in this classification; glucose, sucrose, fructose, maltose, cellulose and starch? Show how oligosaccharides fit into the picture, give an example of one and explain its use or value.
2. Identify and systematically name the isomers of glucose.
3. Give the structure of each isomer of the linear molecules of glucose and fructose. Then, repeat the exercise for the cyclic forms of each compound.
4. Ethanol is being used increasingly as a fuel especially in countries which have no or few reserves of petroleum. This has been true of Brazil in past years although off-shore reserves of petroleum have now been discovered and are being exploited. Sugar from sugarcane plantations is fermented to produce ethanol on an industrial scale, and this is added to petrol or gas as biofuel. Explain why ethanol from this source is regarded as a carbon-neutral biofuel, and give an account of any drawbacks.
5. Give an account of the biochemical processes involved in the conversion of the polysaccharide, cellulose, obtained from straw and wood chippings as by-products in agriculture and forestry, into the biofuel, ethanol.
6. Give an example of a homopolysaccharide containing 3 sugars and a heteropolysaccharide also containing 3 sugars each with 1–4 linkages.

SUGGESTED FURTHER READING

N. G. Clark. 1964. *Modern Organic Chemistry*. Oxford University Press.

R. Cooper and G. Nicola. 2014. *Natural Products Chemistry: Sources, Separations and Structures*. CRC Press, Taylor & Francis Group.

Innovia Films. 2010. *Performance Films for Beverages*, Innovia Limited, UK.

CHINESE CORDYCEPS: WINTER WORM, SUMMER GRASS

Abstract: "Winter Worm, Summer Grass"—the remarkable natural transformation of the mushroom, *Cordyceps sinensis*, which evolves from the caterpillar of a moth!

Natural products chemistry

- Role of the fungus, *C. sinensis*.

Curriculum content

- Chemistry of fermentation
- Redox reactions
- Industrial production of bioethanol.

The Life Cycle of *Cordyceps sinensis*

The *Cordyceps* grows on the Tibetan plateau, which would have been traversed by Marco Polo in following the trail of the Silk Road to the East. He wrote that on resting the yaks, which drew the caravan train, his animals became very frisky. The male yaks began mounting the female yaks in the pack. The yaks had been grazing on the hillsides but he did not realize that they had been eating the plentiful *Cordyceps*! However, today, this mushroom has been collected almost to extinction, yet through Chinese ingenuity has been transformed into a health-giving, energy-boosting food available to the ever growing Chinese population and to the world.

The role of the fungus, *C. sinensis*, is explained in this chapter. Curriculum content relates to the chemistry of fermentation and redox reactions.

The fruiting body appears in summer as brownish-black "blades" which are about 3–6 cm long and are found among grass growing at an altitude of 3,000 m on the Tibetan plateau (Figure 3.19). Apart from provinces of Sichuan and Yunnan in China, the mushroom is also found in Japan, Canada and Russia. The *Cordyceps* are fungi, parasitic upon insect and arthropod larvae. Spores from the fungus infect the larva. Then the spores develop into the thin, thread-like "body" of the fungus, which is called the mycelium. The mycelium consumes the host larva eventually killing and mummifying it. Fungi feed by absorption of nutrients, which the filaments of mycelium find in their hosts. They are nongreen plants without chlorophyll, and cannot photosynthesize their own food.

One particular form of the mushroom, *C. sinensis*, is parasitic on the caterpillars of a moth, which are colonized by the fungus underground. When the host dies, the mycelium of the fungus produces a fruiting body above ground which releases more spores to continue the life cycle (Figure 3.20).

The Perceived Health Benefits of *Cordyceps sinensis*

There are many tales in folklore and in traditional Chinese medicine about the fungus acting as an invigorating tonic. Reports suggest the *Cordyceps* was collected

图 202 冬虫夏草 Cordyceps sinensis（Berk.）Sacc.
（示子座）（293）

FIGURE 3.19 **(See color insert.)** Fruiting body of *C. sinensis*. (R. Cooper collection.)

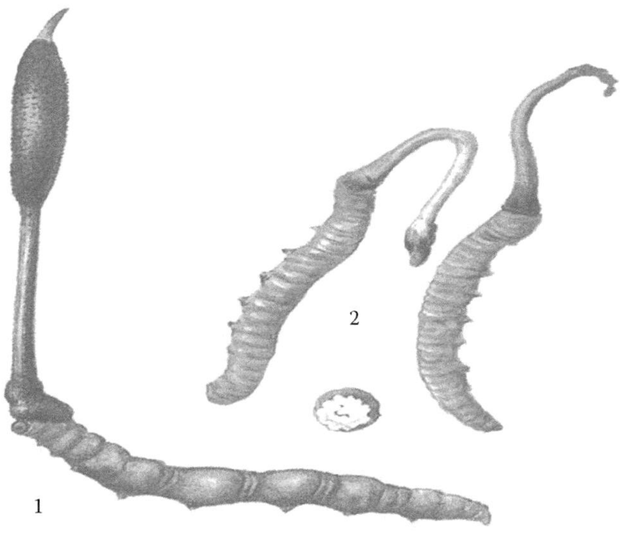

FIGURE 3.20 **(See color insert.)** The fruiting body of *C. sinensis* (dark brown) protrudes from the Earth and grows from the caterpillar of the moth, *Paecilomyces hepiali* Chen (orange). (R. Cooper collection.)

and made into medicinal teas by China's early rulers. Other stories relate that only emperors were given the *Cordyceps* at supper, when it was added to duck soup to give them the energy for nightly pleasures with concubines.

In more recent times, its reputation gained even more prominence when the story broke that the national Chinese coach announced to the world the secret of the caterpillar fungus amid claims of a performance-enhancing nutrient taken by the Olympic Chinese athletes who broke world records at the 1993 China National Games in Beijing.

Modern technology is employed in order to examine whether ancient medicines possess any therapeutic effect. Since the *Cordyceps* is believed to enhance energy and improve performance, athletes can be tested using an ergometer (Figure 3.21). Using the ergometer, it is possible to measure increases in VO_{2max} (oxygen uptake). The idea is to monitor heart rate, pulse and work output in relation to speed and distance on a stationary bicycle. The athlete is connected to a breathing apparatus which measures oxygen intake. Measurements indeed showed that oxygen intake increased when the athlete was taking an extract of the *Cordyceps* and so did stamina too.

The *Cordyceps* has enjoyed immense popularity in the highly populated areas of Eastern China having been brought from areas of Tibet and Western China to the international market. It is hailed as one of China's medical treasures.

In China, the wild fungus is sold as medicine or food. It can be found packaged in small bundles, tied with thread and often attached to the naturally myceliated larvae of the caterpillars. The fungus is eaten in soups, or cooked with meats, and is often administered to elderly patients recovering from illness. This seasonal, ancient Chinese extract has now become a popular fungal-based tonic reputed to address

FIGURE 3.21 An athlete using an ergometer as part of the VO_{2max} test. (Permission granted by D. Jansen, http://www.northeastcycling.com/ne_mail.htm)

many illnesses and conditions. It is claimed that extracts of the fungus have been prescribed for illnesses ranging from headache to Asian flu and to cancer.

The fruiting body of the *C. sinensis* has become the source of one of the most sought-after herbal extracts in the world, and has been collected almost to the point of extinction on the Tibetan plateau.

Most of the world's supply of naturally produced fungus comes from China where an important industry has arisen providing income generation to relieve rural poverty. However, human consumption has been limited due to high price and short supply. Intensive research is being undertaken to generate sustainable supplies and meet spiraling demand. As an alternative, fungal strains from natural *C. sinensis* are isolated in an attempt to achieve large-scale production by fermentation. The fungus is grown in fermentation cultures as pure mycelia in the liquid phase in China, and in the solid state on grains in the western world.

What Is Fermentation?

Fermentation is a process that can occur in nature or in the laboratory. It is a process that converts sugar to alcohol. Fermentation has been used by humans over millennia in the production of many foods and beverages.

Yeast is a single-celled fungus belonging to the phylum, *Ascomycota*, which reproduce by fission or budding. Yeast produces an enzyme which will convert an aqueous solution of sugar (glucose) into carbon dioxide and ethanol in a process known as fermentation.

Carbon dioxide gas bubbles out of the solution into the air leaving a mixture of ethanol and water. Ethanol can be separated from the mixture by fractional distillation.

Yeast is used to make beer and wine. An enzyme in yeast acts on the natural sugars in malt to make beer and grapes to make wine. When the concentration of alcohol reaches about 10%–14% by volume, the yeast dies and fermentation stops naturally, which is why wine is never any stronger in alcohol content, unless it is fortified artificially.

Yeast, of course, is also used in the baking of bread. During baking, the carbon dioxide produced makes the bread rise by creating cavities, and the alcohol evaporates.

Bacteria also promote fermentation. They are used to produce yoghurt and antibiotics, such as penicillin. Fermentation also takes place in active, oxygen-starved muscle cells leaving a residue of lactic acid which causes stiffness in muscle tissue if not flushed out in the bloodstream during a warm-down process after physical exercise.

The famous French microbiologist, Louis Pasteur, is often remembered for his insights into fermentation and its microbial causes. The science of fermentation is known as zymology.

Chemistry of Fermentation: Redox Reactions

Fermentation takes place in the absence of oxygen (when the electron transport chain is unusable) and becomes the cell's primary means of ATP (energy) production. It turns NADH and pyruvate produced in the glycolysis step into NAD^+ and various small molecules.

Nicotinamide adenine dinucleotide (NAD) is an enzyme found in all living cells. NAD exists in two forms, an oxidized and reduced state abbreviated as NAD+ and NADH, respectively.

In the presence of O_2, NADH and pyruvate are used in respiration in an oxidative phosphorylation. It generates a lot more ATP in addition to that created by glycolysis, and for this reason, cells generally benefit from avoiding fermentation when oxygen is available.

The first step, glycolysis, is common to all fermentation pathways:

$$C_6H_{12}O_6 + 2NAD^+ + 2ADP + 2P_i \rightarrow 2CH_3COCOO^- + 2NADH + 2ATP + 2H_2O + 2H^+$$

In this equation, CH_3COCOO^- is called pyruvate, and P_i is phosphate. Two molecules of adenosine diphosphate and two of phosphate (P_i) are converted to two molecules of ATP and two water molecules via a phosphorylation reaction. At the same time, two molecules of NAD are reduced to NADH. These reactions take place in the mitochondria of cells where energy is released from nutrients (see Glossary). The compound is a dinucleotide, consisting of two nucleotides joined through phosphate groups. One nucleotide contains an adenine base and the other is nicotinamide. For more on redox reactions, see Glossary and the section on "Tea: From Legend to Healthy Obsession" in Chapter 4.

Industrial Production of Bioethanol

Fermentation by fungi is used on an industrial scale to produce a number of commercial products such as ethanol, citric acid, steroids and antibiotics. Fermentation must be carried out in the absence of air to make alcohol, otherwise ethanoic acid would be produced instead.

Ethanol can be made from algae by fermentation of the carbohydrates in the starch, which is stored by the plant as a food reserve. The ethanol produced can be used as a biofuel. Also, lipids in oil extracted on a large scale from algae can be made into biodiesel fuel (see the section on "Rediscovering the Traditional Grains of the Americas: Chia and Quinoa" in Chapter 3).

Algae, unlike most green plants, have very thin cell walls with no structural tissue in the form of stems and leaves. Algae are therefore a good natural source of bioethanol because they are high in carbohydrates, which the yeast can use but are low in cellulose.

The industrial process to produce ethanol includes the following stages:

- Growing algae in an aqua culture
- Harvesting the algae as a biomass
- Introducing a yeast to the biomass to cause fermentation
- Extraction of the resultant ethanol from the fermentation solution.

Algae are easy to cultivate in a photobioreactor either to fix CO_2 or to produce biomass. Photosynthesis is performed through chlorophyll in the green algae using sunlight as the energy source. Carbon dioxide is dispersed in aqueous solution in the reactor fluid to make it readily accessible to the algae. The equation describing photosynthesis is

$$6CO_2 + 6H_2O = C_6H_{12}O_6 + 6O_2$$

Heat energy is released, which may be used productively for space heating or to power equipment. Fermentation by yeast of the carbohydrate extracted from the algae biomass releases CO_2, which can be stored and later recycled in the bioreactor to grow more algae, thereby cutting costs.

QUESTIONS

1. Give three examples of redox reactions. Identify donors and acceptors and the changes in oxidation state for each one in your chosen reactions.
2. Ethanol can be made on a large scale for use as a fuel or a solvent from both renewable and nonrenewable resources. Compare and contrast the benefits to society of the production of ethanol using fossil fuels with those arising from processes based on a renewable substrate.
3. Give examples of sources of industrially produced carbon dioxide, and explain how this waste product may be usefully recycled.
4. Fermentation can take place under aerobic or anaerobic conditions. Explain the differences.
5. Why is the *C. sinensis* sometimes referred to as the winter worm, summer grass?

REFERENCE

M. Sutton. 2015. The green molecule. *Chemistry World,* February:50–53.

SUGGESTED FURTHER READING

C. Nappi. 2010. Winter worm summer grass, Cordyceps in crossing colonial historiographies. Digby et al. Eds. Chapter 2. Pages 21–36. Cambridge Scholars Publishing.

Ophiocordyceps sinensis in China. 2010. Edited by the Grassland Monitoring and Management Center, Ministry of Agriculture, ISBN:9877501192144.

GARLIC AND PUNGENT SMELLS

Abstract: Garlic, known as *Allium sativum,* is a species in the onion genus, *Allium.* Garlic has been in use for both culinary and medicinal purposes for over 7,000 years.

Natural products chemistry

- Sulfur compounds in natural products.

Curriculum content

- Organosulfur compounds and their influence upon environmental pollution.

GARLIC

Garlic, known as *A. sativum* (Figure 3.22), is a species in the onion genus. Garlic is frequently used in cooking and in food preparation, but its use is associated with the socially undesirable reputation of "garlic breath." The compound that leads to garlic odor is not present in fresh garlic but is formed only when garlic is crushed

FIGURE 3.22 **(See color insert.)** Garlic bulbs, *A. sativum.* (With permission Steven Foster.)

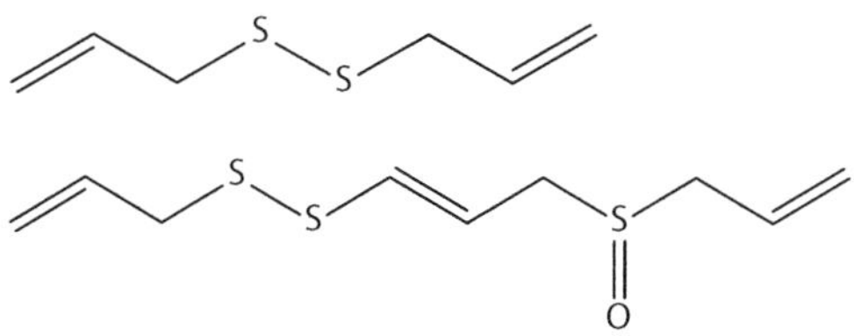

Diallyldisulfide (DADS)

Ajoene

FIGURE 3.23 Two important compounds containing sulfur found in garlic, diallyldisulfide (DADS) and ajoene.

or minced. This action causes enzymes to break down a natural compound named alliin to form allicin, which contributes to the familiar odor of crushed garlic.

Organosulfur Compounds

Organosulfur compounds are common in nature and are often associated with foul smells when chemical breakdown leads to the release of hydrogen sulfide and/or ammonium sulfide.

Indeed, several organosulfur compounds are present in garlic, examples being diallyldisulfide and ajoene (Figure 3.23). Other sulfur-containing organic compounds are diallyl disulfide, allyl methyl sulfide, allyl mercaptan, and allyl methyl disulfide. Of these, allyl methyl sulfide is the compound that takes longest for the body to break down. It is absorbed in the gastrointestinal tract, passes into the bloodstream and moves on to other organs in the body; for excretion through the skin via sweating, via the kidneys in passing urine and is also exhaled from the lungs. Also present in these organosulfur compounds is a functional group known as the allyl group. For more on the allyl function, see lycopene in the section on "Saffron and Carotenoids: Yellow and Orange Dyes" in Chapter 7.

Sulfur-containing compounds are responsible for the antibacterial properties of garlic. These compounds penetrate the membranes of a bacterium cell where they cause changes in the structures of enzymes and proteins, which contain the functional group known as a thiol (–SH), thereby injuring the cell.

Amino Acids Which Contain Sulfur

In the section on "Foods of the Fertile Crescent: Ancient Wheat" in Chapter 3, we introduced essential amino acids. Human beings need 20 essential amino acids for good metabolism and the formation of proteins. Two of these essential amino acids, namely methionine and cysteine, contain a sulfur atom (Figure 3.24). Cysteine is found in garlic and onions.

A molecule of cystine (Figure 3.25) can be produced when two molecules of cysteine interact through the formation of an S–S bond. This type of covalent bond is important in cross-linking protein molecules. The other two ways of cross-linking proteins arise from hydrogen bonding and ionic bonding.

S–S bonds are also prominent in another way in natural products chemistry in that they can be readily involved in redox reactions (see Glossary). One such example

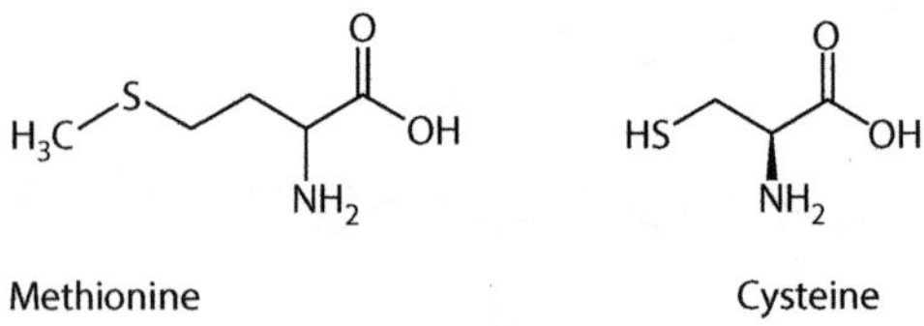

FIGURE 3.24 Two important amino acids which contain sulfur.

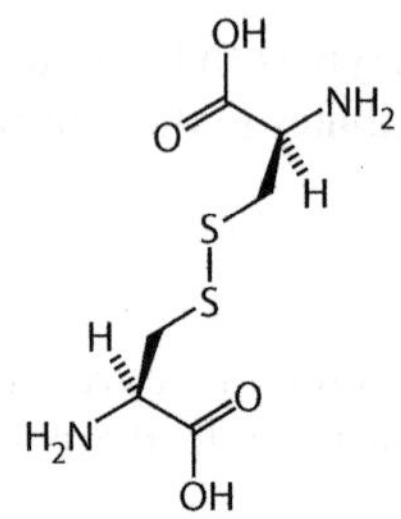

FIGURE 3.25 Cystine.

occurs when human hair is restyled, which is explained in chemical terms in the following manner. Human hair and skin contain approximately 10% of cystine by mass. At the barber's parlor, singed hair characteristically releases some hydrogen sulfide. Also, hair treatments involve the breaking of the sulfur–sulfur bond allowing the hair to be reformed in a different style.

Hair is made mostly of a protein called keratin, which is also present in nails. In hair, keratin molecules are arranged in parallel bundles, which are bound together by cross-linking disulfide bonds. Cysteine, present in one keratin molecule, forms a disulfide bond with the cysteine of a neighboring keratin molecule. In this manner, the greater the number of disulfide linkages present in a strand of hair, the straighter the hair becomes.

Ammonium thioglycolate ($HSCH_2CO_2NH_4$) is a compound that can break disulfide bonds as it contains a thiol functional group. As the disulfide bonds are broken by chemical reduction to thiol groups, the strands of keratin come apart. When hair is restyled with curls, it is called a perm (an abbreviation of permanent waving). Straightening hair is called rebonding. In both cases, the steps are very similar. Once the hair has been washed to clean it thoroughly, ammonium thioglycolate solution is applied for a short while. If a perm is desired, hair is tied around curlers. If the hair is to be rebonded, it is pressed firmly among flat irons until it becomes straight. When the hair has been shaped, the strands of keratin need to be reconnected so that the style is retained. An oxidizing lotion is applied, containing hydrogen peroxide, which reconstitutes the disulfide bonds. Hence, the process involves a redox reaction—also described in the section on "Cocoa: Food of the Gods" in Chapter 4.

Fossil Fuels and Air Pollution

Fossil fuels (coal, petroleum and natural gas) are the decayed remains of ancient organisms and contain, as a consequence, a small proportion of organosulfur compounds. The combustion of these compounds by man leads to an accumulation of sulfur dioxide in the atmosphere, which is a major component of air pollution as the gas combines with water to form droplets of sulfuric acid. These droplets fall as acid rain. Acid rain damages buildings faced in marble or limestone by direct chemical action:

$$H_2SO_4 + CaCO_3 = CaSO_4 + CO_2 + H_2O$$

Sulfur and oxygen are both in Group VI of Mendeleev's classification of the elements, the periodic table. As a consequence of the chemical similarity between sulfur and oxygen, organic compounds containing carbon–sulfur and carbon–oxygen bonds have some similar properties—examples being the alkyl thiols, R-SH, and the alkyl alcohols, R-OH.

QUESTIONS

1. Compare the molecular structures of diallyldisulfide and ajoene presented in the figure earlier in the chapter. Where do you think that the weakest points in the chains of each these molecules occur and why?
2. Give an account of other environmental effects of sulfur dioxide pollution due to increased pH levels in rain and lakes.
3. Describe the measures taken at power stations which use fossil fuels to prevent sulfur dioxide from entering the atmosphere.
4. Apart from power stations, what other sources of air pollution by sulfur dioxide are there and what are the preventive measures that may be taken?
5. Sulfur and oxygen are both in Group VI of Mendeleev's classification of the elements, the periodic table. Explain why sulfur and oxygen are found in the same chemical group in the periodic table.
6. As a consequence of the chemical similarity between sulfur and oxygen, organic compounds containing carbon–sulfur and carbon–oxygen bonds have some similar properties. Compare and contrast an alkyl alcohol, R-OH, with an alkyl thiol, R-SH.

SUGGESTED FURTHER READING

H. P. Koch, L. D. Lawson, Eds. 1996. *Garlic. The Science and Therapeutic Application of Allium sativum and Related Species*, 2nd edition. Williams & Wilkins.

4 Beverages

INTRODUCTION

Folk medicine was commonly characterized by the application of simple indigenous remedies while other botanical preparations were sold as teas and spices. These plant extracts were known within their respective communities and prepared locally. Some plant materials were used in liquid concentrates while others were presented as dried, ground-up powders. In more recent times, examination of extracts from these preparations has led to the identification of many chemical constituents, which has fostered development of modern phytomedicines and herbal drinks.

Early humans lived near rivers, springs and lakes in order to ensure an adequate supply of fresh water. The addition of taste to water has developed over time and drinks have continued to shape human history ever since. The drinking of beverages is associated with a variety of roles, including acts of ceremonial purpose and celebration, as an indication of status and power and as a vehicle to imbibe a health-saving medicine.

Some beverages have become popular at different times, in various places and cultures, and influenced the course of history. During the fifteenth century, China, the great tea producer of the time, imposed a policy of "strength through isolation" and stopped international trade and exploration. In contrast, at the same time, Western rulers were witnessing and experiencing tremendous changes. The Renaissance, The Reformation and the growth of nation states stimulated great interest in exploration of new opportunities for trade throughout the world. A clear dichotomy in outlook had arisen between East and West. China relied on tradition and isolation at a time when Europe was fascinated by new developments in warfare, technology, economics and political organization. This would prove to be disastrous for China in the centuries to come. Eventually, Portugal was the first country to set up trade with Asian countries, and other European countries soon followed. In 1600, Queen Elizabeth granted a charter to the John Company (eventually to be known as the British East India Company) for the sole purpose of promoting trade with Asia. Tea was first brought to London in 1657 as a medicinal herb and was originally supplied only by apothecaries.

Well-known and popular, nonalcoholic beverages are described: coffee, tea and cacao, each of which contains the stimulant, caffeine. They have, in common, an important role in history and trade right up to the present day. We also include a beverage that is less well known called maca, which grows in Peru at elevations above 4,000 m. Maca is also used traditionally as a food and to enhance fertility in human beings and domesticated animals.

TEA: FROM LEGEND TO HEALTHY OBSESSION![*]

Abstract: Tea: from legend to healthy obsession. The history of tea as a popular beverage from Asia to the rest of the world! Tea has a history that is inspiring and disheartening, uplifting and disturbing. Tea, with all of its associated rituals and ceremonies, has been the inspiration for art and poetry for centuries but has also been the object of power and manipulation bringing both great pleasure and great pain to millions of people for many hundreds of years. Today, new scientific advances suggest important medicinal benefits and healing properties arise from drinking infusions of green and black teas.

Natural products chemistry

- Catechins.

Curriculum content

- Phenolic ring
- Electrophilic and nucleophilic substitution reactions.

HISTORICAL NOTE ON THE ORIGINS OF TEA

All true tea comes from a single species of plants, *Camellia sinensis* (this includes black, green, oolong and white tea). Other drinks that are popularly called "tea" but do not come from this plant, such as rooibos, chamomile, mint, etc., are more accurately called tisanes. Scientists have divided the tea species into two distinct varieties, *C. sinensis var. sinensis*, which is indigenous to and has been cultivated in western China for nearly 2,000 years, and *C. sinensis var. assam*. The latter is a virtual newcomer to the world, as it was only discovered in the Assam region of India in the nineteenth century. This variety is indigenous to a large geographic region, including India, Myanmar (Burma), Thailand, Laos, Cambodia, Vietnam and southern China.

No one really knows when the leaves of *C. sinensis* were first used to make tea. A legend dating to 3,000 BC, tells us that the mythical Chinese emperor, Shen Nung, was the first to taste tea. He was considered the father of Traditional Chinese Medicine and was said to have tasted and tested thousands of herbs to determine their possible usefulness. According to the tea legend, Shen Nung and his followers had stopped to rest under a small tree before continuing with their journey. The emperor was warming a pot of water on a fire when a leaf from the tree fell into the water. Shen Nung drank it and was said to have immediately recognized the health benefits of the plant. Archaeological evidence actually predates this legend and suggests that tea was first consumed during the early Palaeolithic period (about 5,000 years ago).

[*] Published in part by L. C. Martin, R. Cooper. 2011. From herbs to medicines: A world history of tea—From legend to healthy obsession. *Alternative and Complementary Therapies* 17:162–168.

Tea was originally used as a medicine and was considered so effective that, by the fourth century, it was an important part of Chinese life, and was used to cure a diversity of ailments including poor eyesight, fatigue, rheumatic pains, and problems with kidneys and lungs. Because successful processing methods had not yet been discovered, tea in the early years was a bitter brew, masked with a variety of additives, including onions, ginger, salt, and oranges.

The art of processing tea evolved from using the raw leaf to baking it into a dried brick that could be carried great distances and used over a long time. The new processing method also resulted in a dramatically improved taste. Tea became so popular that people throughout China, from peasants to the Imperial Court, drank this beverage daily. The T'ang Dynasty was well known for valuing the arts, poetry, landscape gardening and music, all of which were influenced by the spread of tea. Owing to the great increase in popularity of the drink, teahouses and tea gardens sprang up in cities and towns throughout the empire.

During the Song (Sung) dynasty, (~927 AD), tea was one of China's most important trade items, and a trade route called the Tea and Horse Caravan Road was developed between Tibet and the tea-growing regions of China. There, valuable horses from Tibet were traded for vast quantities of tea, and records show that in 1 year, 20,000 warhorses from Tibet were exchanged for up to 34 million pounds of tea.

TEA AND OPIUM

China was the only country in the world producing tea at that time. Yet, China was insular, with distrust of foreigners and a difficult trading partner. Inspired by the fortunes to be made in the tea trade, the British were relentless and became obsessed with the idea of finding a place where tea could be cultivated successfully. Both solutions came from the same place—India.

The British East India Company established the first trading post on the coast of western India in 1619, where they seized thousands of acres of land in southern India and began growing poppies for opium, which the Chinese merchants would trade for. In spite of Chinese government bans, in 1830, 2.5 million pounds of opium were traded for tea, and addiction to the drug became rampant throughout China.

This tension resulted in the Opium Wars, fought between the British and the Chinese, and wars were fought in 1840 and again in 1856, which left the British as victors on both occasions. The resulting treaties benefited British traders who were passionate about the idea of growing tea in India and, by the end of the century, more than half a million acres in India were planted with tea.

TEA SPREADS THROUGHOUT THE WORLD

In early times, although China was the only place where tea was cultivated for export, its importance in other countries grew quickly. By the sixth century, drinking tea was a part of daily life throughout Korea and Japan. In the twelfth century, the Japanese Samurais came into power and by this time, tea masters had developed even better ways of processing the tea leaves. Instead of being baked into a brick, the leaves were dried and powdered, then boiling water was added and the brew was whipped with a bamboo whisk until foamy. The taste was so superior that it was sometimes called "frothy jade" and is known today as Matcha tea. This has a "grassy" sweet taste, and although delicious by the standards of the twelfth century, the beverage cannot compare to the vast diversity of tastes and flavors that came with the discovery of new tea processing methods by the sixteenth century.

A new Chinese method of processing, drying the leaves and then steeping them in hot water, brought out the delicate and sweet flavor of tea, and eventually the British—and others became obsessed with the taste of tea. By 1675, tea could be found in food stores in England and, by the end of the seventeenth century, both black and green teas were being shipped to England from China in great quantities. In 1734, Thomas Twining opened the first teahouse in London and others soon followed. By the end of that century, tea had become an essential part of British life when it was estimated that workers and labourers spent approximately 10% of their salaries on tea and sugar. The British had an insatiable thirst for tea. Drinking tea became a social occasion. Late afternoon tea became an established custom throughout the country, first among the upper class and royalty, but eventually spreading to the peasant and lower classes as well.

The Growing and Processing of Tea

All true tea, green black and white, is derived from *C. sinensis*, an evergreen shrub of the Theaceae family. Unlike black tea, which has undergone some form of oxidation, green tea is harvested and carefully dried with very little further process. Green tea is mainly consumed in the form of a brewed beverage. Successful tea cultivation requires moist humid climates provided most ideally by the slopes of Northern India, Sri Lanka, Tibet and Southern China. Green tea is consumed predominantly in China, Japan, India and a number of countries in North Africa and the Middle East, whereas black tea is consumed predominantly in Western and some Asian countries, and often taken together with milk.

After the young fresh leaves (white or green leaves) are picked, they undergo one or more of the following processes. The process chosen determines whether or not the tea will remain green, or in the form of oolong or black tea (Figure 4.1).

Withering: Fresh, green leaves and buds are softened by withering. The leaves are allowed to air dry in the sun or are placed on racks in a large, heated room. During this stage, the starch in the leaves begins to convert to sugar. For white or green tea, the leaves are withered for 4–5 hours—for oolong and black tea, almost twice as long.

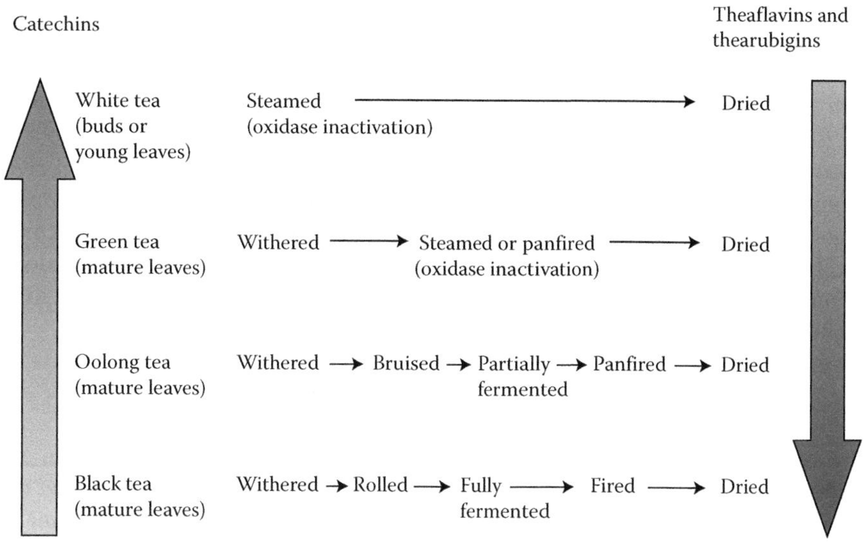

FIGURE 4.1 Schematic of the tea extraction processes.

Rolling: After withering, the leaves are rolled in a process, originally performed by hand but in modern times by machine, that twists and crushes the leaves. This action releases the sap and exposes it to the air which stimulates oxidation.

Oxidation: For white and green tea leaves, this step is missed out. Black tea is fully oxidized. Oolong tea is partially oxidized. Rolled leaves are placed on trays and left in a cool, damp place for 1–3 hours, and the leaves turn from green to a copper color.

Drying: After the rolled leaves are oxidized (in the case of black tea), they are dried with hot air to quickly stop any further oxidation, also called "fermentation," or the growth of mould. The leaves are then sorted and packaged.

Tea as a Modern Medicine

Historically, the medicinal use of green tea dates back to China 4,700 years ago. Today, drinking tea continues to be regarded in Asia as a generally healthful practice. Numerous scientific publications relating to clinical and epidemiologic studies now attest to the health benefits of both black and green teas. Although all tea contains beneficial antioxidants, high quality green and white teas have greater concentrations than black tea. Today, scientists believe that the main active ingredients of green tea include catechins and theanine which are considered to have properties, which help in the fight against cancer.

The benefits of drinking green tea are not limited to cancer but also thought to extend to cardiovascular disease. The presence of catechins in black tea may reduce cholesterol levels in blood and are potentially useful compounds in the reduction of cholesterol in serum.

The history of tea is rich, complicated and fascinating. Perhaps as much as any plant on earth, tea has changed the world we know. Tea has been consumed throughout the world for almost 2,000 years: first beginning in the use of tea as a medicinal herb in China; the incomparable formal influence of the tea ceremony in Japan; the British passions for tea as a beverage and as a social catalyst. As the modern world continues to shrink into a global community, tea will continue to wield civilizing influence. Although the popularity of tea comes with a price as the impact on the environment in tea-growing regions is significant, the calming and soothing effects of tea are benefits, which cannot be ignored. As long as people can pause long enough to brew a "cuppa," there is a good chance that we will remember the finer points of civilization and appreciate better the world around us.

Green Tea Helps against Cancer

Research studies on the health benefits of drinking tea, particularly green tea, are producing encouraging results in cancer research. The studies in both Asia and the West seem to indicate that drinking green tea contributes to the fight against many different kinds of cancer including those of the stomach, esophagus, the ovaries and the large intestine or colon. From epidemiologic studies, an empirical link between green tea and cancer-prevention properties was made in the late 1980s. These studies showed that the onset of cancer in patients in Japan who had consumed 10 cups of green tea per day was 8.7 years later among females and 3 years later among males, compared with patients who had consumed less than 3 cups per day. A possible correlation between large consumption of green tea and low incidence of prostate and breast cancer in Asian countries, where green tea consumption is high, has been postulated. However, because of the complicating influence of many other variables in lifestyle inherent in such a study, a definitive link between green tea and these beneficial effects on cancer cannot, as yet, be concluded. Results are, however, more promising in the use of polyphenols in green tea, such as catechins, to combat skin cancer. Polyphenols from green tea are thought to help prevent the onset and growth of skin tumors (melanomas) through their effectiveness as a natural sun screen in absorbing harmful UVB radiation.

Catechins: Key Phenolic Constituents in Green Tea

Catechin molecules present in tea contain four hydroxyl groups attached to two benzene rings which, therefore, have strong phenolic properties. Catechin is a flavanol (Figure 4.2), a natural phenol and an antioxidant, which belongs to the chemical family known as the flavonoids (see also the sections on "Cocoa: Food of the Gods" and "Reversible Colors in Flowers, Berries and Fruit").

Catechin possesses two benzene rings (called the A- and B-rings) and a nonaromatic cycle (the C-ring) with a hydroxyl group on carbon 3. There are two chiral centers in the molecule on carbons 2 and 3, and so there are four diastereoisomers (see Glossary). The two isomers which are in the *trans* configuration are called *catechin*, while the other two isomers in the *cis* configuration are known as *epicatechin* (see also the section on "Pacific Yew Tree" in Chapter 2, for *trans* and *cis* isomerism).

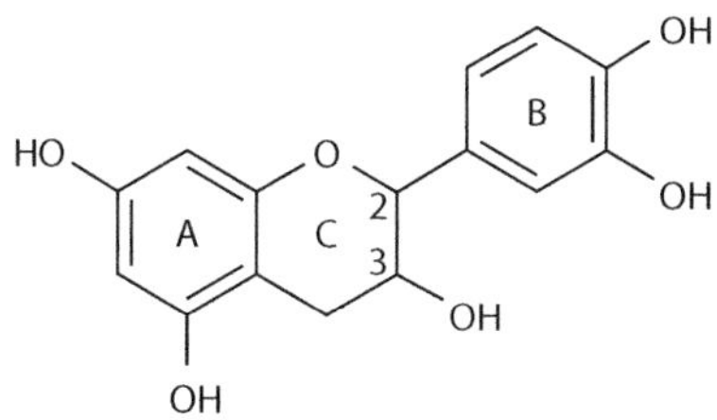

FIGURE 4.2 Catechin.

The main active constituents of green tea include polyphenolic compound which are derivatives of catechin; epicatechin-3-gallate, epigallocatechin and epigallocatechin-3-gallate. Each of these catechins may be responsible for the anticarcinogenic and antimutagenic activities of green tea. Other polyphenols in green tea include flavanols, caffeine and small amounts of ethylxanthines.

Polyphenols in green tea have been shown to be powerful antioxidants with anticarcinogenic properties. Human studies on the pharmacokinetics of polyphenols in green tea have been conducted. The evidence suggests that ingested polyphenols and their metabolites may play a role in the action against gastrointestinal cancers.

Many *in vitro* and *in vivo* studies demonstrate that polyphenols from green tea are anticarcinogenic by inducing apoptosis (see Glossary). Probable action mechanisms depend upon the antioxidant properties of polyphenols; an ability to scavenge free radicals and stimulation of detoxification systems through selective induction or modification of favorable metabolic enzymes.

The Properties of Phenol and Phenols

Phenol, also known as carbolic acid, is an aromatic organic compound with the molecular formula C_6H_5OH having a phenyl group ($-C_6H_5$) bonded to a hydroxyl group (–OH) (see also the section on "Cacao (Cocoa): Food of the Gods"). Phenol and its chemical derivatives (phenols) are key building blocks in the commercial production of polycarbonate materials, nylon, detergents, herbicides and numerous pharmaceutical drugs.

Phenol is also a building block of many natural products. The hydroxyl group in phenol may be replaced, for example, by a methyl or acetyl group, or by a carboxyl group, or by an ether linkage.

Phenols Compared with Alcohols

Although at first glance they may appear similar to alcohols, compounds based on phenol have unique, distinguishing properties. Unlike in alcohols where the hydroxyl group is bound to a saturated carbon atom, it is significant that in phenol the hydroxyl group is attached to an unsaturated benzene ring. Phenol is quite soluble in water where partial dissociation takes place to give an acidic solution involving a negative phenoxide ion and a positive hydrogen ion. An explanation for the acidity of phenol is resonance stabilization by the aromatic ring of the phenoxide anion. Negative charge is delocalized over the benzene ring and the oxygen atom.

Electrophilic Substitution Reactions of Phenol

However, ease of electrophilic substitution (see Glossary) in the ortho and/or para positions (or the 2 and 4 carbon atoms) of the aromatic ring is a notable chemical property of phenol. Phenol will react with electrophilic substances such as nitric acid or concentrated sulfuric acid at room temperature to give in the latter case a mixture of ortho and para sulfonic acids. When, in turn, ortho and para sulfonic acids are fused with potassium hydroxide at 350°C the corresponding dihydric phenols, catechol and quinol (hydroquinone), are produced. Their systematic formulae are $C_6H_4(OH)_2$ (1, 2) and $C_6H_4(OH)_2$ (1, 4), respectively. In this reaction, strongly nucleophilic hydroxyl anions substitute for the sulfonyl groups in the aromatic ring.

Interaction with Light in the Visible Part of the Spectrum of Electromagnetic Radiation

Light in the visible spectrum can interact with the delocalized electrons in the phenoxide ring which gives rise to the variety of pigmentation in plants containing phenolic compounds. These aspects are explored more fully in the section titled "Colorful Chemistry: A Natural Palette of Plant Dyes and Pigments" in Chapter 7.

QUESTIONS

1. Both phenol and ethanol contain the OH group. Describe reactions in which both substances behave similarly to one another and other reactions in which they behave quite differently.
2. Explain how water is important as a solvent in fostering partial dissociation of phenol.
3. Phenol is a building block of many natural products. The hydroxyl group in phenol may be replaced, for example, by a methyl or acetyl group, or by a carboxyl group, or by an ether linkage.

   However, ease of electrophilic substitution in the ortho and/or para positions (or the 2 and 4 carbon atoms) of the aromatic ring is a notable chemical property of phenol. Describe a theoretical model which adequately accounts for these properties of phenol.
4. Explain why concentrated rather than dilute sulfuric acid is necessary to allow electrophilic substitution in the benzene ring to take place.
5. Phenol is a tremendously important chemical used as a feedstock in the organic chemicals industry. Trichlorophenol™, commonly known as TCP, is a derivative of phenol which has medical applications due to its antiseptic properties. Describe three other diverse applications of products derived from phenol in the modern world.

REFERENCE

L. C. Martin, R. Cooper. 2011. From herbs to medicines: A world history of tea—From legend to healthy obsession. *Alternative and Complementary Therapies* 17:162–168.

SUGGESTED FURTHER READING

J. Blofeld. 1985. *The Chinese Art of Tea*. Shambhala.
J. K. Fairbank, M. Goldman. 1998. *China: A New History*. Belknap Press.
A. MacFarlane, I. MacFarlane. 2003. *The Empire of Tea: The Remarkable History of the Plant That Took Over the World*. The Overlook Press.
L. C. Martin. 2007. *Tea: The Drink that Changed the World*. Tuttle Publishing.
R. Moxham. 2003. *Tea: Addiction, Exploitation, and Empire*. Carroll & Graf.
W. Ukers. 1935. *All about Tea*. Tea and Coffee Trade Journal Co.

COCOA (CACAO): FOOD OF THE GODS*

Abstract: Cocoa—food of the gods! Originating in South America, and developed into the sacred beverage of Aztec and Mayan tradition, products of this modest tree of the Amazon forests led to worldwide passion for chocolate and cocoa. Recent emerging scientific data add medicinal importance to its value as a pleasant beverage. As well as being tasty, chocolate contains compounds known as flavonoids, which have been shown to have a wide range of pharmacological benefits and produce color in many plants.

Natural products chemistry

- Flavonoids (polyphenols)
- Color arising from phenolic compounds.

Curriculum content

- Properties of phenol
- Free radicals and antioxidants.

Origins

Everyone loves chocolate! But where does it come from? It was once prized by Aztec warriors, and today by millions of people around the world. Furthermore, it may even be good for your health. Chocolate originates from cocoa. In 2006–2007, over 1.2 million tons of cocoa were produced in the largest producing country of Cote d'Ivoire (Ivory Coast). Today, cocoa consumption ranges from 0.1 kg/person/year in China to 11 kg/person/year in Ireland, with the United States in the middle of this range at 5 kg/person/year.

The chocolate, or cocoa tree, is commonly known as cacao to botanists. The scientific name of the chocolate tree (*Theobroma cacao*) literally means "food of the gods." European interest can be traced to the early 1500s, when Columbus engaged natives in the Gulf of Honduras, who gave him *xocoatl* made of cacao, honey, spices and vanilla. When sweetened with sugar, chocolate became popular throughout Europe. It was offered in fashionable drinking houses and valued for its alleged aphrodisiac properties.*

Cacao and the Aztecs

The Aztecs named the cacao *xocolatyl* from which the word chocolate is derived. The Mayans and Aztecs drank their cocoa in a bitter concoction made with chili peppers, corn mash, vanillin and other spices. The Mayan drink "xocoatl" means

* Published in part by R. Cooper and T. J. Gianfagna. 2012. Cocoa: Food and medicine of the gods. *Alternative and Complementary Therapies* 18(2): 84–90.

"bitter water." Drinking a cocoa beverage before a long march or expedition was believed to increase energy and stamina.

Processing Cacao Beans

Cacao beans (Figure 4.3) come from the pods found on cocoa trees growing in warm, humid places near the Equator.

The main producing areas, Ghana, the Ivory Coast, Brazil and Nigeria are all perfect locations, and production is increasing in Malaysia too.

Typically, cocoa seeds are fermented, dried and roasted during their processing into chocolate. For more on fermentation, see also the section on "Chinese Cordyceps: Winter Worm, Summer Grass" in Chapter 3. Fermentation occurs in large piles of cocoa beans using the natural yeasts and bacteria that are present to produce some of the flavor precursors, which we associate with chocolate. This processing creates a thick paste called cocoa liquor. Combination with cocoa butter (the fat component) and sugar creates dark chocolate and, when milk is added, milk chocolate. Chocolate contains fiber (most of which is lost with processing); minerals such as magnesium, copper and iron (providing a significant portion of the recommended daily allowance); the monosaturated fatty acid, oleic acid and saturated fatty acids, mainly palmitic acid and stearic acid.

FIGURE 4.3 Cocoa beans. (With permission Savor AnnArbor, http://www.savorannarbor.com/product/chocolate-heaven)

HISTORICAL NOTE: FIRST CHOCOLATE WAS SOLD AS A MEDICINE

In 1687, an English doctor and botanist, Sir Hans Sloane, was traveling in Jamaica where he tried their local chocolate drink, which was improved by adding milk, and this recipe was brought to England and first sold as a medicine. As more trade emerged, chocolate became increasingly popular, and to meet the new demand, cocoa plantations were developed in the West Indies, the Far East and Africa. However, the eating of chocolate was not in vogue until early Victorian times, and then gradually spread across Europe as it became fashionable with the European royalty, the wealthy and the nobility. However, by the end of the nineteenth century, milk was added leading to today's popular chocolate products.

Cocoa and Cardiac Health

In a very specific example of comparing yesteryear's and today's preparations, we can observe two groups of Kuna Indians. A very low incidence of hypertension has been reported in Kuna Indian groups living on islands off the coast of Panama who consume large amounts of unprocessed cocoa everyday containing high flavonol content. The Kunas live an "idyllic" island life and have much lower rates of cardiovascular diseases (hypertension, cancer and diabetes) than populations of Kuna Indians who have migrated to the mainland of Panama. Even though one may think that living on a tropical island would reduce blood pressure and stress levels, it appears that the Kunas on the mainland are as equally satisfied with their lives as the Kuna islanders. One explanation to account for differences in their health is that the Kuna islanders drink a minimally processed cocoa beverage throughout the day, whereas the Kuna people on the mainland drink a cocoa beverage that would resemble the more commonly and highly processed cocoa beverage available everywhere else.

Cocoa and Diabetes

The island population of Kuna Indians was also noted to have a lower incidence of type-2 diabetes. This disease results when the body no longer responds effectively to the insulin produced by the pancreas. The most important function of insulin is to direct the body to remove glucose from the blood and either store it in adipose (fat) tissue, store it in the liver for future use or send it to muscle cells to provide energy for activity. Could there be a connection between cocoa consumption and the prevention of diabetes as well as cardiovascular diseases? There is now good evidence of a correlation between hypertension and insulin resistance. Insulin resistance refers to an inability to reduce blood glucose levels (hyperglycaemia) despite adequate production of insulin by the pancreas. Chronically high levels of glucose in the blood lead to oxidative stress, inflammation and eventually diabetes and cardiovascular disease. Insulin increases blood flow to microcapillaries in skeletal muscle, liver and adipose tissues, and promotes cell surface receptors in these tissues to take up glucose.

Few would disagree that the seeds of this tropical tree, native to the upper Amazon rainforest, produce one of the most sought after food products, both today and in antiquity. Moreover, cocoa contains natural compounds that may reduce the incidence of diabetes and cardiovascular diseases. If, indeed, it can be shown that cacao helps prevent the insulin resistance and hypertension which lead to cardiovascular disease and diabetes, then *Theobroma cacao* is truly both food and medicine of the gods!

Chemical Constituents of Cocoa

Cocoa seeds contain flavonoids which are polyphenols. Polyphenols are powerful antioxidants which protect cells from damage by free radicals and the resultant inflammation of tissues.

Free Radicals and Antioxidants

A free radical is an atom, molecule or ion that has unpaired valence electrons. This feature makes free radicals highly reactive toward other substances. Most free radicals are reasonably stable only at very low concentrations in inert media (such as a noble gas) or in a vacuum.

Examples of free radicals are the hydroxyl radical (HO•), a molecule that is one hydrogen atom short of a water molecule, and the oxygen atom which is an intermediary in the formation of ozone in the upper atmosphere when ultraviolet light dissociates an oxygen molecule:

$$O_2 = 2O; \quad O_2 + O = O_3$$

Free radicals may be created in a number of ways, including synthesis with very dilute or rarefied reagents, reactions at very low temperatures, or the breakup of larger molecules. The latter can be brought about by ionizing radiation, heat, electrical discharges, electrolysis and chemical reactions.

Free radicals are intermediate stages in many chemical reactions. Free radicals and other oxygen-derived species are constantly generated in the body both by accident and during metabolic processes. The reactivity of different free radicals varies but some can cause severe damage to important biological molecules such as DNA (see Glossary), or those which form major organs such as the liver. Antioxidants offer some defense as they react with and minimize the formation of oxygen-derived free radicals. Consequently, antioxidants derived from diet may be particularly important in helping us to stay healthier for longer.

Flavonoids (or Polyphenols)

Flavonoids (polyphenols) also impart a purple color to the coating of the bean but are largely lost during fermentation, roasting and other processes leading eventually to the production of chocolate. Flavonoids are widely distributed in plants, fulfilling many functions. Flavonoids are the most important plant pigments for flower coloration, producing red, purple or blue pigmentation in petals to attract pollinators (see also the section on "Colorful Chemistry: A Natural Palette of Plant Dyes and Pigments" in Chapter 7 and, for more on flavonoids in particular, the section on "Reversible Colors in Flowers, Berries and Fruit" also in Chapter 7). In higher plants, flavonoids are also involved in the symbiotic fixation of nitrogen from the air.

Most chocolate drinks contain about 25 mg of flavonoids. Chemical extracts reveal among them the presence of two common chemicals; theobromine and caffeine.

It is interesting to note that modern nutritional supplements, containing cocoa flavonoids or pure epicatechin, are often used by runners and body builders who try to gain the same advantages, which the Aztecs sought from the foamy cocoa drink. In the eighteenth century, concoctions were made by apothecaries and chemists who considered their cocoa blends as a kind of medicine. However, these forms of beverage were very different from today's chocolate. They contained several flavonoids, which are typically lost in today's processing and had no added sugars or milk. Food processing does affect the level of flavonoids remaining in the finished product and levels are widely variable between types of milk, dark and white chocolate and cocoa products.

QUESTIONS

1. The main producing areas of cocoa beans, Ghana, the Ivory Coast, Brazil and Nigeria, are all perfect locations, and production is increasing in Malaysia too. What are the similarities and growing conditions needed in these different countries?
2. The process of fermentation has been introduced in the section on "Chinese Cordyceps: Winter Worm, Summer Grass" in Chapter 3 and has been reinforced in this chapter on the processing of cocoa beans. Describe the process of fermentation in your own words.
3. Chemical extracts from cocoa reveal the presence of two alkaloids, theobromine and caffeine. Draw their chemical structures. Why are they considered as alkaloids?
4. What other well known plant do you think contains theobromine and caffeine?
5. In a very specific example, a very low incidence of hypertension has been reported in Kuna Indian groups living on islands off the coast of Panama, who consume large amounts of unprocessed cocoa with high flavonol content everyday. Why is monitoring hypertension important to our health?
6. Explain the difference between Type 1 and Type 2 diabetes.
7. What is insulin and what is its function in the human body?

REFERENCE

R. Cooper and T. J. Gianfagna. 2012. Cocoa—Food and medicine of the Gods. *Alternative and Complementary Therapies* 18(2):84–90.

SUGGESTED FURTHER READING

S. D. Coe, M. D. Coe. 1996. *The True History of Chocolate.* Thames & Hudson.

L. E. Grivetti, H.-Y. Shapiro, Eds. 2009. *Chocolate. History, Culture and Heritage.* John Wiley & Sons, Inc.

COFFEE: WAKE UP AND SMELL THE AROMA!

Abstract: Wake up and smell the aroma! With origins in remote regions of East Africa, coffee is now a global mega crop. The widely consumed beverage is increasingly recognized not only as a stimulant but also for its health benefits.

Natural products chemistry

- Caffeine
- Nitrogen in the organic ring
- Cyclic aromatic amines.

Curriculum content

- Cyclic aromatic amines
- Zwitterions
- The process of decaffeination of coffee.

HISTORICAL NOTES

COFFEE FROM THE ARAB WORLD ONWARD

In early times, use of coffee remained largely confined to Ethiopia where its native beans were first cultivated, particularly in the Ethiopian highlands. The spread of coffee can be traced from the Yemen northward to Mecca and Medina and then to the larger cities of Cairo, Damascus, Baghdad and Istanbul. The first coffee house opened in Istanbul in 1471. As the Arab world began expanding its trade horizons, the coffee beans moved into northern Africa and large-scale cultivation began. Eventually, the beans entered Indian and European markets as the popularity of coffee drinking spread to Italy and the rest of Europe.

Coffee was first imported to Europe through Venice. Trade wars were now on! The race among Europeans to secure coffee trees or fertile beans was eventually won by the Dutch, who in the late seventeenth century brought back coffee plants for growing in greenhouses. Largely through the efforts of the British East India Company, coffee became available in England no later than the sixteenth century. The first coffee house in England was opened in St. Michael's Alley in Cornhill, London. In France, the most important introduction of coffee occurred in 1669, when Soleiman Agha, Ambassador from Sultan Mehmed IV, arrived in Paris with his entourage bringing with him a large quantity of coffee beans. Not only did they provide their French and European guests with coffee to drink, they managed to firmly establish the custom of drinking coffee. The first coffee house in Austria opened in Vienna in 1683, after the Battle of Vienna, by using supplies obtained as spoils from the defeat of the Turks. Importantly, their coffee houses helped popularize the custom of adding sugar and milk to the coffee.

The introduction of coffee to North America can be attributed to French colonization of many parts of the continent and the West Indies. The first French coffee plantations were founded in Martinique in the Caribbean in 1720, eventually enabling the spread of coffee cultivation to Haiti, Mexico and other islands of the Caribbean.

COFFEE AND SLAVERY

When coffee entered the Caribbean region in the early eighteenth century, it flourished, in no small part, due to slave labor on the plantations. It is estimated that from the sixteenth to nineteenth centuries, over 1 million slaves were transported from Africa to Cuba in order to work these crops. Although the production and selling of sugar in the country began the slave holding, the presence of coffee played an equally important role in establishing the slave trade in Cuba. When coffee reached Cuba, farmers immediately welcomed it; coffee required (a) less land to grow and (b) little machinery to tend and process it. The territory now known as Haiti supplied half the world's coffee in the middle of the eighteenth century, however, terrible working conditions led to uprisings, and Haiti's coffee industry never fully recovered.

The first coffee plantation in Brazil was established in 1727. The plant produced smaller beans and was deemed a different variety of Arabica known as var. Bourbon. Cultivation began after independence in 1822 with heavy reliance on slave labor from Africa. Further, large tracts of rainforest were cleared for coffee plantations. However, by the 1800s, coffee went from an elite indulgence to a drink for the masses. In the nineteenth and early twentieth centuries, Brazil became the biggest producer of coffee. However, competitive pricing by other nations, such as Colombia, Guatemala, Indonesia and Viet Nam, has affected Brazil's monopoly.

Finally, although the origins of coffee cultivation began in Ethiopia, only small amounts were exported until the twentieth century. Once coffee beans from Brazil were reintroduced to Africa, plantations began to thrive in Kenya and Tanzania—not far from the place of origin: Ethiopia, 600 years earlier.

Early Use

The Kefficho people, living in the region known as Keffa in Ethiopia, are believed to be among the first to discover and recognize the energizing effect of an extract from coffee beans. The use of coffee is believed to have spread from Ethiopia to Egypt and Yemen. The earliest credible evidence of either coffee drinking or knowledge of the coffee bush appears in the middle of the fifteenth century in the Sufi monasteries of Yemen. By the sixteenth century, knowledge of coffee had reached the rest of the Middle East, Persia, Turkey and northern Africa.

An entertaining story involves a shepherd, who noticed the energizing effects when his goats nibbled on the bright red berries of a certain bush. He chewed on the fruit and his exhilaration prompted him to offer these berries to the Muslim holy men at a nearby monastery. At first there was disapproval of their use and they threw them into the fire, generating an enticing aroma. In their curiosity, they raked the roasted beans from the embers, ground them up, and dissolved them in hot water. Hey presto—the world's first cup of coffee!

Coffee and Caffeine

Coffee comes from the *Coffea* genus of flowering plants. Although caffeine is not responsible for the well-known aroma of coffee, caffeine is present in the seeds where it protects the seeds as a secondary metabolite (see Glossary) due to its toxicity to herbivores. One of the most popular of coffee-producing plants is *Coffea arabica* (Figure 4.4).

The chemical formula of caffeine (Figure 4.5) is $C_8H_{10}N_4O_2$. It is weakly basic, and is a white colorless powder when in its anhydrous state. The solubility of caffeine is 2 g/100 mL in water at room temperature, which increases significantly to 66 g/100 mL when it is added to boiling water.

FIGURE 4.4 **(See color insert.)** *Coffea arabica.* (With permission from S. Foster.)

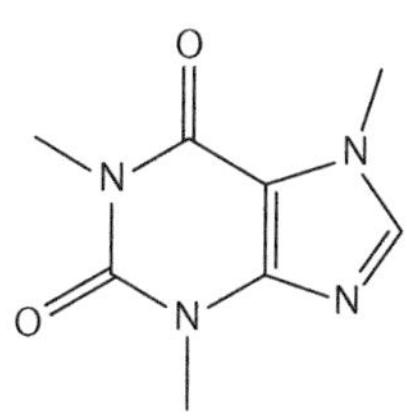

FIGURE 4.5 Structure of caffeine.

Zwitterions

A zwitterion has both a positive and a negative charge at different positions within its structure. In other words, it is a dipolar ion. Amino acids (see the section on "Foods of the Fertile Crescent: Ancient Wheat" in Chapter 3) are well known for this property in that the molecules possess both basic and acidic functional groups in the form of an amine and a carboxylic acid. The six-member ring of caffeine contains two amide functional groups, which exist in resonance producing an intramolecular separation of charge as in a zwitterion (see the diagrammatic representation of the structure in Figure 4.6). The possibility of hydrogen bonding when caffeine is added to a polar solvent such as water helps to explain the moderate solubility of the compound.

Thus, in a similar manner, we can see the two resonant molecular structures for caffeine. The position of the equilibrium is related to the pH of the solution containing caffeine (Figure 4.7).

Cyclic Aromatic Amines

A molecule of caffeine clearly possesses four nitrogen atoms in two joined, heterocyclic rings, which are aromatic in nature. Two of the nitrogen atoms are in the configuration of an amide, whilst the other two are formally in the relationship of a tertiary amine.

The chemistry of amines and amides is presented and reinforced in the section on "Tobacco: A Profound Impact on the World" in Chapter 5. As amines are ultimately derivatives of ammonia, they contain a lone pair of electrons located on the nitrogen atom, which makes them basic compounds. However, the degree of basicity is markedly influenced by neighboring atoms and whether or not the nitrogen atom

$CH_3CH\ (NH_2)\ CO_2H$ <=> $CH_3CH\ (NH_3)^{(+)}\ CO_2^{(-)}$

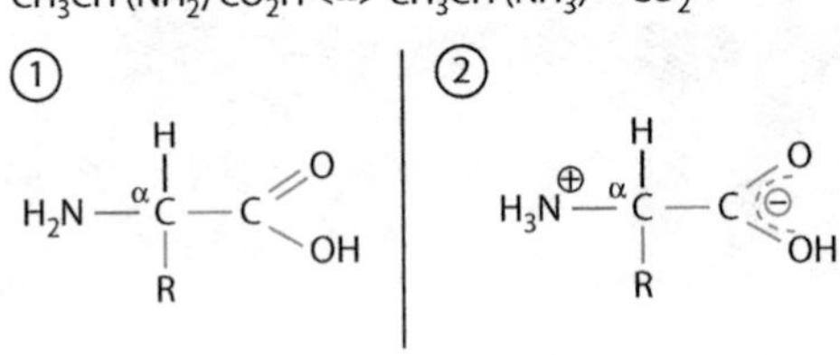

FIGURE 4.6 The zwitter ion forms of an amino acid.

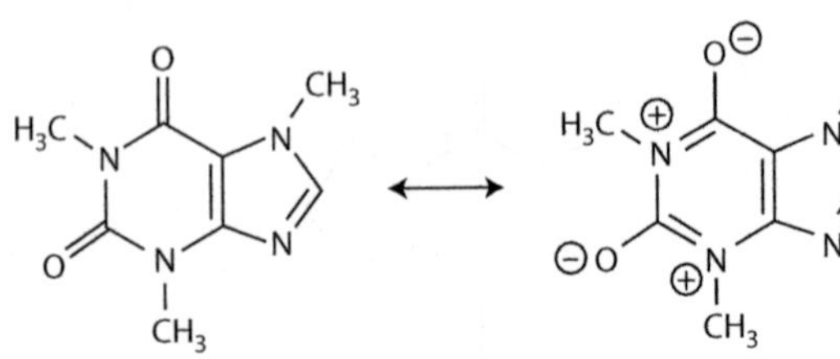

FIGURE 4.7 Resonant molecular structures of caffeine.

is incorporated into a heterocyclic, aromatic carbon ring as shown for caffeine. The aromatic structure decreases the basic properties of the compound due to the delocalization of electrons. This is an important point, which governs the fact that caffeine is only weakly basic. However, this property significantly affects the process of removal (or reduction in the level) of caffeine in beverages such as coffee and tea in the decaffeination process. As caffeine is weakly basic, it will dissolve in polar solvents such as water or ethanol.

HISTORY OF DECAFFEINATION

The first successful decaffeination was achieved in 1820 when the German chemist, Runge, analyzed the constituents of coffee to discover a possible link between drinking coffee and insomnia. There was a more significant breakthrough by Ludwig Roselius in 1903. He pretreated coffee beans with steam which eventually became the basis for commercial production of decaffeinated coffee in the early twentieth century.

The Process of Decaffeination of Coffee

There are several ways to remove caffeine from coffee, and three methods are presented here.

Extraction Procedure I: Solvent Extraction Using Water

Runge developed a commercial procedure which depends upon the solubility of caffeine in water that can be reproduced in the laboratory. Thus, clean coffee beans are first soaked in water. Caffeine dissolves along with some other compounds in low concentrations. The solution is then passed over charcoal. The charcoal absorbs and retains all but the molecules of caffeine, which can be recovered from the aqueous phase by standard laboratory techniques. The process is repeated with fresh charcoal. Eventually, the coffee beans are dried and are caffeine free.

Extraction Procedure II: Solvent Extraction Using Dichloromethane (DCM)

First, ground coffee in aqueous sodium carbonate is refluxed for 20 minutes; the mixture is filtered and cooled. The aqueous filtrate is partitioned into DCM and repeated several times to extract more caffeine. The addition of sodium carbonate, a weak base, converts the protonated form of caffeine, which is naturally present in coffee, to its normal, free caffeine form (Figure 4.8).

Extraction Procedure III: Supercritical Carbon Dioxide Extraction

There are two clear advantages to this method: elimination of the use of a flammable, toxic solvent, and the caffeine is more easily removed from the final product. However, sophisticated equipment is required to sustain the high pressure and temperature required to maintain carbon dioxide in a supercritical fluid state (abbreviation, SC-CO_2; also see the Glossary for definition of the critical point).

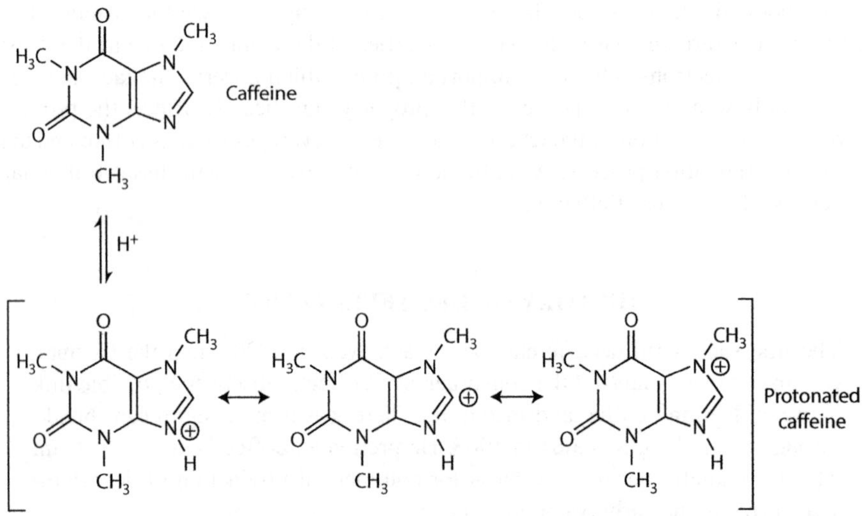

FIGURE 4.8 Addition of sodium carbonate converts the protonated form of caffeine to its free form.

Supercritical carbon dioxide is an excellent nonpolar solvent for caffeine. In the extraction process, CO_2 is forced through the ground coffee beans at temperatures above 31°C and high pressures above 73 atmospheres. Under these conditions, CO_2 is in a supercritical fluid state when it has the properties of both a gas, allowing the solvent to penetrate deep into the beans, but also the properties of a liquid, whereby it can dissolve 97%–99% of the caffeine. The solution of caffeine in supercritical SC-CO_2 is then sprayed with high-pressure water to remove the caffeine. The caffeine can be removed from aqueous solution by charcoal and refined by distillation, or recrystallization if required.

SC-CO_2, as a nonpolar solvent, dissolves nonpolar solutes from the mixture. The subsequent addition of a more polar solvent, water, dissolves the somewhat polar solute, caffeine. The careful selection and use of cosolvents in this way to partition a complex mixture enriches the extraction of a target solute (in this example caffeine) from a complex natural product without damaging the latter.

Multiple partitioning in a continuous industrial extraction process removes most of the caffeine. Coffee beans enter at the top of an extractor vessel with fresh CO_2 entering at the bottom. Recovery of caffeine is achieved with water in a separate absorption chamber. The process benefits from a pretreatment step. The material, first, is soaked with ultrapure water when hydrogen bonds linking caffeine to its natural matrix are ruptured. Swelling and bursting of the cell membrane also enhances diffusion of the solutes into the solvents.

The extract from the beans contains the compounds besides caffeine, which contribute to the flavor of coffee. Finally, the extract is dried leaving decaffeinated coffee with its original flavor intact. The quality of recovered caffeine produced can reach a purity of greater than 94% making it acceptable for use in the soft-drinks and pharmaceutical industries.

QUESTIONS

1. Which of the four nitrogen atoms in the caffeine molecule do you expect to be the most basic? Explain in terms of the delocalization of electrons over the two fused rings showing the possible different resonant structures of the molecule.
2. Explain fully in terms of molecular structure and chemical properties why caffeine, as an aromatic organic compound, is soluble in water at all.
3. Give a full account of how water is used in a supercritical fluid state in the industrial-scale generation of electrical power.
4. Previously, organic solvents such as hexane, benzene, chloroform and carbon tetrachloride have been typical choices for the extraction of a solute by partitioning. Explain why the use of ethyl benzoate for this purpose is much more preferable in the laboratory.

SUGGESTED FURTHER READING

S. C. Chew. 1974. *The Crescent and the Rose*. Oxford University Press.

R. Cooper, G. Nicola. 2014. *Natural Products Chemistry: Sources, Separations and Structures*. CRC Press, Taylor & Francis.

M. Pendergrast. 1999. *Uncommon Grounds: The History of Coffee and How It Transformed Our World*. Basic Books.

The Blessed Bean—History of Coffee. Encyclopedia Britannica. 1954. Otis, McAllister & Co.

MACA FROM THE HIGH ANDES IN SOUTH AMERICA

Abstract: How did the Inca—healthy people with high reproductive rates—prosper at high altitude in the Andes? The answer may be the ancient staple food, looking like a little radish and known as maca, which is obtained from the plant, *Lepidium meyenii.*

Natural products chemistry

- Alkaloids
- Indole as a building block in nature
- Introducing indole alkaloids.

Curriculum content

- Comparison of the basic properties of secondary amines located in a heterocyclic ring with aromatic heterocyclic compounds containing nitrogen.
- A simple practical technique—acid–base extraction.

The Maca Plant

Maca (*Lepidium meyenii*) grows in Peru at elevations over 4,000 m. It is used traditionally as a food and to enhance fertility in both human beings and domesticated animals. The environment in which the maca grows in Peru would be exceptionally challenging for most plants; intense cold, intense sunlight and strong drying winds. It is a staple food and has been cultivated in the Andes for 1,500–2,000 years, and may have contributed to the survival of the healthy indigenous populations with high reproductive rates in the high Andes (Figure 4.9).

Maca as a Beverage and a Food

Once unknown in the Western world, maca is now being recognized well beyond the Andean region. The useful part of maca is found in the root. Maca can be consumed

FIGURE 4.9 The radish-like root of the maca.

as a beverage. Juice is recovered from the root by boiling in water or by extraction with alcohol.

In foods, maca is eaten either baked or roasted and may also be prepared as a soup. Maca flour is being sold in health food stores as a more healthy option than conventional flour. In an increasingly interdependent world, cultural fusion in culinary practice has led to products such as spaghetti being made from maca flour.

The Medicinal Value of Maca

Traditionally, maca has been taken orally to relieve symptoms of anemia and chronic fatigue as it has the capability to enhance stamina, athletic performance and memory. Extracts of maca are also used to treat female hormone imbalance and menstrual irregularities and for enhancing fertility.

A study of the health status of adults, 35–75 years old, in the Peruvian central Andes involved comparison of those who used maca with those who did not. The majority of people (80%) in the sample used maca solely for nutrition. Maca was clearly associated with better health indicators; stronger bones (fewer fractures), reduced incidence of chronic mountain sickness, lower body mass index and lower systolic blood pressure.

Chemical Composition of Maca

The dried maca root contains carbohydrates (60%); protein (10%); fiber (8.5%) and lipids (2%) including linolenic acid, palmitic acid and oleic acid. It contains significant amounts of minerals including iron, calcium, copper, zinc and potassium. Importantly, there are secondary metabolites in trace quantities, which are unique to maca and may be used to identify the plant from samples. These include the alkaloids; macaridine, macaene and macamides.

Generally, alkaloids (see Glossary) do have biological effects on humans and animals. Alkaloids are a large group of naturally occurring chemical compounds that are produced by a wide variety of organisms including bacteria, fungi, plants and animals. Alkaloids always contain at least one basic nitrogen atom.

The tuber of the maca plant contains the alkaloid, (1*R*, 3S)-1-methyltetrahydrocarboline-3-carboxylic acid, also known as a carboline. It is an indole alkaloid which is reported to exert influence on the human central nervous system. Its chemical structure is shown in Figure 4.10.

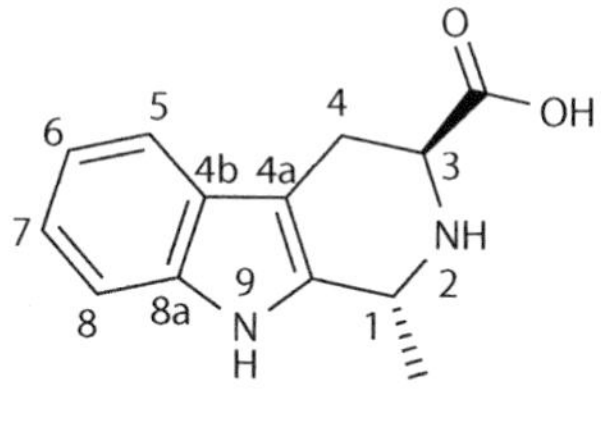

FIGURE 4.10 (1*R*, 3S)-1-methyltetrahydro-carboline-3-carboxylic acid.

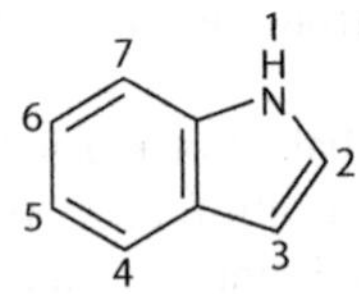

FIGURE 4.11 Indole structure showing the convention of numbering the carbon atoms.

Indole

Indole is an important building block of many naturally occurring chemicals (see also the sections on "Africa's Gift to the World" and "Woad"). The indole building block is that group of atoms forming two fused rings on the left of the molecular diagram (see Figure 4.11). Indole is an aromatic, heterocyclic organic compound. It has a bicyclic structure consisting of a six-membered benzene ring fused to a five-membered pyrrole ring. Pyrrole contains a nitrogen atom and for more on amines, amides and pyrrole refer to the section on "Tobacco."

Scientific interest in indole intensified when it is was realized that the indole building block is present in many important alkaloids.

Pure indole is a solid at room temperature. Indole is widely distributed in the natural environment and can be produced by a variety of bacteria which are single-celled plants. Bacteria in the human gut produce indole as a degradation product of the amino acid, tryptophan, and therefore, indole occurs naturally in human feces producing that intense fecal odor.

The molecular structure of tryptophan contains the indole building block. The molecule is completed by an amino acid, derived from propionic acid, which is attached to the indole building block at the second carbon atom from the nitrogen atom (see Figure 3.10 for the molecular structure of tryptophan in the section "Foods of the Fertile Crescent: Ancient Wheat" in Chapter 3). Though tryptophan is essential for human life, significantly, it cannot be synthesized in the body and, therefore, must be part of our diet. Essential amino acids, such as tryptophan, have been reviewed earlier in the section on "Foods of the Fertile Crescent: Ancient Wheat." Essential amino acids act as building blocks in the biosynthesis of proteins. Fortunately for the human race, tryptophan is a routine constituent in many foods, being plentiful in bananas, dates, milk products, chocolate, meat, fish, poultry, eggs, oats and peanuts.

Key Chemical Properties of Indole

Electrophilic Substitution

By far the most reactive position on the aromatic indole molecule for electrophilic substitution is the C-3 carbon atom on the pyrrole ring, which is much more reactive than a carbon atom within a benzene ring.

It follows then that since the pyrrole ring is by far the most reactive part of a molecule of indole, electrophilic substitution of the carbocyclic or benzene ring can only take place once the N-1, C-2 and C-3 positions have been substituted.

Much more on electrophilic substitution is presented in the section on "Tea: From Legend to Healthy Obsession" in this chapter.

Indole and Bases

Initial appearances can be deceptive as indole is quite unlike an amine. In fact, indole is hardly basic at all. This is entirely due to the fact that indole is an aromatic compound in which electron delocalization of the electron pair of the nitrogen atom over and below the plane of the two rings plays a large part in stabilizing the molecule. This situation is quite comparable to the properties of the similar heterocyclic aromatic compounds, pyrrole and pyridine. Consequently, only a very strong mineral acid, such as hydrochloric acid, is able to protonate the nitrogen atom of indole.

Acid–Base Extraction

Alkaloids are produced in nature by a large variety of organisms including bacteria, fungi, plants and animals. Alkaloids are a group of naturally occurring chemical compounds that almost always contain at least one basic nitrogen atom. They can be purified from crude extracts by acid–base extraction.

The theory behind acid–base extraction is that salts, which are ionic, tend to be water soluble, while neutral molecules tend not to be. The addition of a mineral acid to a mixture of an organic base and acid will result in the acidic functional group remaining uncharged, while the base will be protonated. If the organic acid, such as a carboxylic acid, is sufficiently strong, its self-ionization can be suppressed by the additional acid.

Conversely, the addition of a base to a mixture of an organic acid and base will result in the base remaining uncharged, while the acid loses hydrogen ions to give the corresponding salt. Once again, the self-ionization of a strong base is suppressed by the added base.

The acid–base extraction procedure can also be used to separate very-weak acids from stronger acids and very-weak bases from stronger bases.

Usually, the mixture is dissolved in a suitable solvent, such as dichloromethane or diethyl ether (ether), and is poured into a separating funnel. An aqueous solution of the acid or base is added, and the pH of the aqueous phase is adjusted to bring the compound of interest into its required form. After shaking and allowing for phase separation, the phase containing the compound of interest is collected. The procedure is then repeated with this phase at the opposite pH range. The process can be repeated to increase the separation. It is often convenient to have the compound dissolved in the organic phase after the last step so that the evaporation of the solvent yields the product.

Acid–base extraction is also covered in the sections on "Africa's Gift to the World: The Madagascan Periwinkle" and "Coca and Cocaine."

Purification of Indole Alkaloids by Acid–Base Extraction

Since the molecules of indole alkaloids contain at least one, weakly basic, nitrogen atom, the indole alkaloids can be purified from crude plant extracts by acid–base extraction.

The carbolines present in the maca plant are indole alkaloids, which are formed by plants through a condensation reaction involving amino acids and reducing sugars (aldoses). Significantly for the purposes of acid–base extraction, carboline has the functional group of a secondary amine at the number 2 carbon atom.

Indole Alkaloids

The indole alkaloids form one of the largest classes of alkaloids, with more than 4,100 different compounds known. There are two types of indole alkaloids—those which also include an isoprene building block and those which do not. For example, the indole alkaloid found in maca is an indole alkaloid without a terpenoid group.

The physiological action of certain indole alkaloids on human beings, animals and birds has been well known down the ages. The group of indole alkaloids containing terpenoids has some interesting members described below!

Strychnine

The French chemist Pelletier is revered as the founding father of alkaloid chemistry. The indole terpenoid alkaloid, strychnine (Figure 4.12), was isolated as early as 1818 by Pelletier and Caventou.

They used extracts from the plants of the *Strychnus* genus as source material. *Strychnus* is a flowering tropical plant belonging to the family, Loganiaceae or Strychnaceae, which also includes trees and lianas. The roots, stems and leaves of these plants are well known to indigenous peoples as sources of poisonous compounds, such as strychnine and curare, which they used efficiently in hunting (discussed below and also in the "Importance and Role of Natural Products" within the "Introduction" in Chapter 1).

The strychnine molecule has a number of functional groups, one of which, the ether link, is not treated extensively in this book, primarily because it is relatively uncommon in naturally occurring compounds and is not very reactive, given the stability of the carbon–oxygen single bond. This subject is also treated in the section titled "Attacking Malaria: A South American Treasure (but Not Gold) and a Chinese Miracle." The relative stability of the carbon–oxygen single bond is in marked contrast to the reactivity of the carbon–oxygen double bond, which is somewhat strained. The latter functional group features in the extensive organic chemistry of

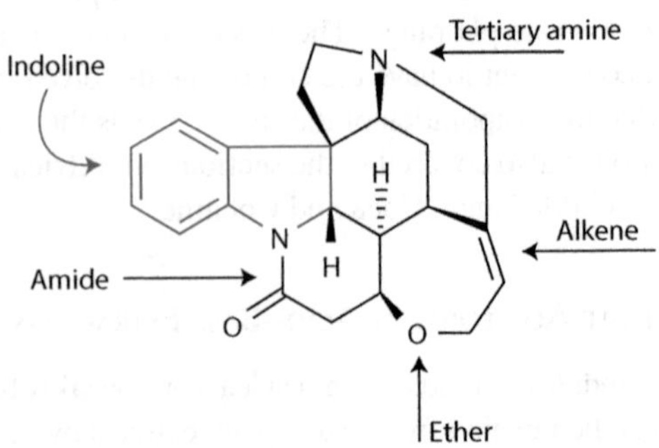

FIGURE 4.12 Molecular structure of strychnine showing the indole building block.

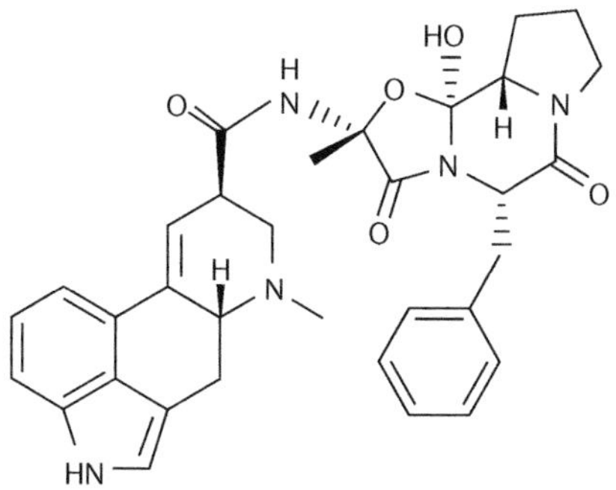

FIGURE 4.13 Structure of ergotamine.

the aldehydes, ketones, carboxylic acids and esters covered extensively in the sections on "Asian Staple: Rice," "A Plant from the East Indies, Camphor," "A Steroid in Your Garden," "Morphine: A Two-Edged Sword" and "Europe Solves a Headache."

Ergotamine

It has been recognized for a long time that human consumption of contaminated grains of the cereal, rye, affected by the ergot fungus, which is parasitic upon it, causes death by poisoning. The active chemical, an indole alkaloid called ergotamine (Figure 4.13), was isolated from extracts in 1918.

Lysergic Acid Diethylamide (LSD)

Lysergic acid diethylamide, known in every-day language as LSD (Figure 4.14), was first synthesized from naturally occurring ergotamine in 1938. LSD is a prohibited recreational substance in many countries of the world, as it has pronounced psychedelic properties. LSD is a large molecule made up from two of the most common building blocks found in nature—an indole and a terpene.

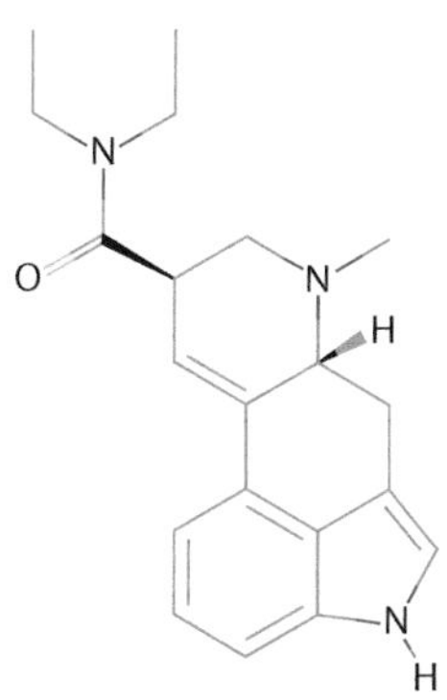

FIGURE 4.14 Structure of LSD.

Curare

The word "curare" comes from the South American Indian word, *ourare*, meaning arrow poison. Curare has been a poison used with arrows or darts by South American natives to hunt animals. They are able to eat the animals subsequently, without any untoward effects. The active extract, tubocurarine, was first isolated in London by H. J. King in 1935 from a sample obtained from the large liana, *Chondodendum tomentosum*, native to the rain forest in South America. He also identified the structure of the molecule.

Tubocurarine is a toxic alkaloid and skeletal muscle relaxant, known as a long-duration antagonist for the nicotinic acetylcholine receptor. In biology, a chemical compound, such as acetylcholine, which binds to a receptor and activates the receptor producing a biological response, is given a special name. The compound is known as an agonist. A compound such as tubocurarine which blocks this process is known as an antagonist.

Acetylcholine is an organic substance which functions as the neurotransmitter activating neuroreceptors at neuromuscular junctions.

You can see from Figure 4.15 that acetylcholine is an ester formed from acetic acid and an amino alcohol called choline. Acetylcholine is present in the peripheral nervous systems of animals where it activates the neurons, which control the movement of muscles, notably those attached to the skeleton. Acetylcholine is also known to occur in the brain where it is believed to influence learning, memory and mood, and could be linked with the memory difficulties experienced by sufferers of Alzheimer's disease.

A neuroreceptor found in the muscle tissues of animals is nicotinic acetylcholine, which is activated by acetylcholine. The binding of the acetylcholine ion to that of the neuroreceptor is reversible. The "firing" of the neurons attached to muscle tissues causes the muscles to contract followed by relaxation when the reaction is reversed.

However, the tubocurarine ion in curare can also bind to either of these neuroreceptors, but does so irreversibly and, therefore, causes paralysis in muscle tissue. The diaphragm, which evacuates the lungs of air-breathing animals is a large muscle which is also affected as curare from the tip of the dart is carried around the bloodstream. In effect, the animal dies from suffocation.

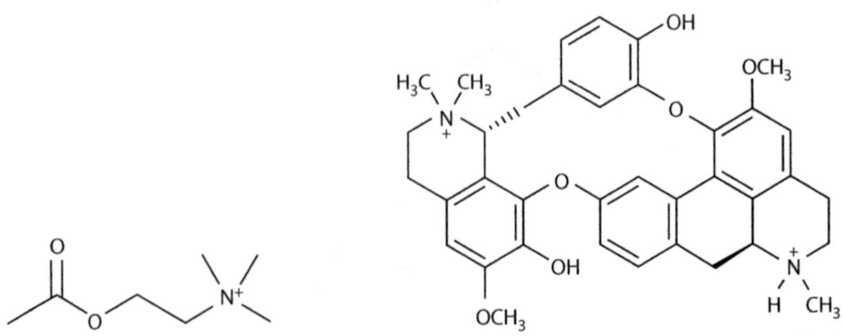

FIGURE 4.15 Acetylcholine, a neurotransmitter, and tubocurarine, a component of curare.

The nicotinic neuroreceptors present in muscle tissues are very large molecules, measuring 290 kDa, and have a tubular shape with five openings at one end. Electron-rich, amine functional groups are exposed in this open structure through which ligands can be formed by binding with specific other molecules or ions that are electrophilic, such as acetylcholine or tubocurarine. More on the chemical binding of ligands is to be found in the section titled "Our World of Green Plants: Human Survival" in Chapter 7.

Even a cursory comparison of the molecular structures of acetylcholine and curare reveals why the poison is so effective. Both cations have a clear similarity in that amine functional groups are present, which have been protonated. Tubocurarine blocks the access of acetylcholine by binding to the neuroreceptor.

QUESTIONS

1. What functional groups and common naturally occurring chemical building blocks are present in (1*R*, 3S)-1-methyltetrahydro-carboline-3-carboxylic acid? Describe the chemical properties you would expect the compound to have.
2. Explain why the substance, (1*R*, 3S)-1-methyltetrahydro-carboline-3-carboxylic acid, may be separated from crude chemical samples of the Maca plant by acid–base extraction.
3. Explain fully the reasons why the nitrogen atom in the indole building block is not basic.
4. Identify all of the functional groups present in strychnine, and describe the range of organic chemistry which you expect the compound might show.

REFERENCES

G. F. Gonzales. 2012. Ethno-biology and ethno-pharmacology *of Lepidium meyenii* (maca), a plant from the Peruvian Highlands. *Evidence-Based Complementary Alternative Medicine* 2012 Article ID 193496.

H. J. King. 1935. Extraction of tubocurarine from the liana, *Chondodendum tomentosum*, and identification of the structure of the molecule. *Chemical Society* 57:1381 and *Nature, London* 135:469.

S. Piacente, V. Carbone, A. Plaza, A. Zampelli, C. Pizza. 2002. Investigation of the tuber constituents of maca (*Lepidium meyenii*). *Journal of Agricultural and Food Chemistry* 50(20):5621–5625.

R. B. Van Order, H. G. Lindwall. 1942. Indole. *Chemical Reviews* 30:69–96.

SUGGESTED FURTHER READING

M. E. Moseley. 2001. *The Incas and Their Forebears: The Archaeology of Peru*. Thames & Hudson Publishers.

National Research Council. 1989. *Lost Crops of the Incas: Little-Known Plants of the Andes with Promise for Worldwide Cultivation*. Page 57. National Academy Press.

D. Seigler. 2001. *Plant Secondary Metabolism*. Springer Publishers.

5 Euphorics

INTRODUCTION

Certain drugs are known as euphorics because of their impact on the human brain. Plants containing substances, which generate euphoria have played an important role in the culture of humankind. As source plants and their extracts were gradually "discovered" by explorers, trade in "new" drugs became embedded in the global economy and linked inextricably with colonial expansion.

Examples of controlled euphoric drugs are opium from the poppy plant, cocaine from the coca leaf, and marijuana from cannabis. If taken properly under medical supervision, these drugs are of benefit to humankind. However, we wish to make it clear that we do not condone their use for recreational purposes.

There are also "soft" drugs, such as nicotine in tobacco and caffeine in tea and coffee, which do have a degree of social acceptance.

SUGGESTED FURTHER READING

M. Jay. 2011. *Emperors of Dreams. Drugs in the Nineteenth Century*. Dedalus.

MORPHINE: A TWO-EDGED SWORD*

Abstract: The poppy has been traded through Asia and Arabia for centuries and continues to have an impact on underground commerce and societies. The poppy yields one of the most important and effective narcotics used in medicine today. Morphine is a legal drug used by medical doctors to relieve severe or agonizing pain and suffering. The drug acts directly on the central nervous system.

Natural products chemistry

- Morphine and codeine
- Heroin alkaloids.

Curriculum content

- Formation of esters from acids and alcohols
- Acetylation of morphine to form codeine and heroin.

JOE AND MIKE: THE TWO-EDGED SWORD

Everyday, our local hospital receives many patients. Patients requiring critical attention are always admitted firstly through the emergency ward. On one evening this summer, two teenagers arrived separately, both requiring immediate care but for different medical needs. Although each one was treated for different conditions, you will soon discover that their stories are very much linked.

The first teenager, whose name was Joe, had a serious bike accident, and it appeared that his leg was broken. He was feeling so much pain, the doctors needed to administer by injection several small doses of a pain-killing drug called morphine. After the surgery to set his bone in place, he recovered well and the doctor carefully prescribed some more morphine to keep him comfortable and reduce the pain. Even though Joe felt better, and was grateful for all the pain killers, he worried whether he could become addicted to the morphine, which he had been receiving for the past couple of days. The doctors assured him that the likelihood of this was extremely low and that he would make a complete recovery. A week later, Joe left the hospital with a big cast on his leg—ready for his school friends to sign all of their get-well messages.

The second teenager was Mike who entered the emergency room a few hours later. He was found comatose. The doctor soon discovered he was suffering from an overdose of drugs—in this case heroin—but the doctors and nurses were slowly able to revive him. When he began to recover, Mike told the doctor how he had started on recreational drugs and began to smoke heroin. Then he learned how to inject himself with the heroin with much stronger effect on his body and mind. Mike explained to the doctor how he enjoyed it at first, but became addicted, and continually craved for more. Without the drug, he felt miserable. However, to continue buying the drugs,

---

* Published in part by R. Cooper. 2013. Morphine & heroin: The Yin & Yang of narcotics. *ACS ChemMatters* Dec:14–16.

he needed to steal money and do other illegal things. The doctor was sympathetic and told Mike that withdrawal from this powerful drug is a long and painful process, needing much medical patience, time and support, but there was good chance of recovery, if Mike was willing to work with the doctors and did not take any more heroin. Not all addicts are able to "stay clean." For Mike, that night in the emergency ward became his first long step on the road to recovery.

Morphine, opium and heroin are all related chemicals, and all have the high potential for addiction. Tolerance and psychological dependence develop rapidly, although physiological dependence may take several months to develop.

Morphine and opium are both classified as a narcotic—a drug to dull the senses. This means they have pain-killing properties and euphoric and hallucinatory effects. Morphine deadens pain, produces elation, induces sleep and reduces stress. Opium induces gentle, subtle, dream-like hallucinations very different from the fierce and unpredictable weirdness of the psychedelic drug known as LSD described in the section on "Maca from the high Andes in South America" in Chapter 4.

A SHORT HISTORY OF OPIUM

Ancient peoples either ate parts of the poppy flower or converted them into liquids to drink. By the seventh century, the Turkish and Islamic cultures of western Asia had discovered that the most powerful medicinal effects could be obtained by igniting and smoking the poppy's congealed juices, and the habit spread. The widespread use of opium in China dates to tobacco-smoking in pipes introduced by the Dutch from Java in the seventeenth century. Whereas the Javanese ordinarily ate opium, the Chinese smoked it. The Chinese mixed opium with tobacco, two products traded by the Dutch. Pipe-smoking was adopted throughout the region which resulted in increased opium-smoking, both with and without tobacco.

The great renaissance scientist, Paracelsus (1490–1541), concocted laudanum ("something to be praised") by extracting opium into brandy (ethanol), thus producing, in effect, a tincture of morphine. Laudanum was historically used to treat a variety of ailments, but its principal use was as an analgesic and cough suppressant. Laudanum can be habit forming. As their opioid tolerance increased, so did users' consumption of tinctures. By the nineteenth century, vials of laudanum and raw opium were freely available at any English pharmacy or grocery store. Until the early twentieth century, laudanum was sold without a prescription and was a constituent of many patent medicines. During these times, opium was viewed as a medicine, not a drug of abuse. The chemists and physicians most actively investigating the properties of opium were also its dedicated consumers; and this may have colored their judgment. Today, laudanum is recognized as addictive and is strictly regulated and controlled throughout most of the world.

In 1907, the British phased out India's opium export to China. At the time, they realized the social consequences of heroin, which is a derivative of opium, and they had lost out to Bayer in heroin production, and China began poppy field eradication. Even as treaties were enacted calling, for countries to ban the illegal trade of opium, criminal syndicates and illegal trafficking of opium became prevalent. However, the use of opium for medicinal use is legal.

In North America, the initial history of poppy was somewhat more peaceful. During the first few centuries of European settlement, opium poppies were widely cultivated, and early settlers dissolved the resin in whisky to relieve coughs, aches and pains.

As well as being used for opium production, poppy seeds were used as food. The plant produces many small black seeds and these are used as a common garnish on rolls. Poppy seeds can also be ground into flour, used in salad dressings, added to sauces as flavoring or thickening agents, and the oil can be expressed and used in cooking. Poppy heads are infused to make a traditional sedative drink.

In the eighteenth to nineteenth centuries medical doctors had long hunted for effective ways to administer drugs without ingesting them. Taken orally, opium is liable to cause unpleasant gastric side effects. The development of the hypodermic syringe in the middle of the nineteenth century allowed the injection of pure morphine. Morphine use became rampant in the United States after its extensive use by injured soldiers on both sides of the Civil War. In late nineteenth-century United States, opiates were cheap, legal and abundant. Only when morphine addiction became well understood at the beginning of the twentieth century were regulations imposed for its withdrawal as an over-the-counter medicine.

Production of Morphine in the Natural World

Morphine is a chemical found in nature, specifically in the poppy plant (*Papaver somniferum*) (Figure 5.1). The plant is typically found growing in arid climates, for example in Afghanistan, within a temperature range of 7–23°C, and where the pH of the rich moist soil is 4.5–8.

Morphine (Figure 5.2) is the most abundant alkaloid found in opium. Opium is a complex mixture of chemicals from the poppy plant, which contains sugars, proteins, fats, wax, latex, gums, water, ammonia, lactic acid and numerous alkaloids, most notably morphine. The content of morphine in the plant is generally 8%–17% of the dry weight of opium, although specially bred cultivars reach 26%. Opium poppy contains at least 50 different alkaloids, but most of them are in very low concentration. Morphine is the main alkaloid. Yet, all alkaloids, including morphine, can be purified initially from crude extracts by the acid–base extraction technique summarized in the section on "Maca from the High Andes in South America" in Chapter 4, and in the section on the Madagascan periwinkle entitled "Africa's Gift to the World" in Chapter 2.

The harvesting of opium, even in modern times, is carried out in the traditional way. Farmers harvest opium 2 weeks after the petals fall from the unripe seed case. The seed case remains on the stem and 2 or 3 vertical slots are made into the skin

FIGURE 5.1 **(See color insert.)** Poppy, *Papaver somniferum,* showing the petals and pods. (With permission from S. Foster.)

using a sharp blade. An incision, if skin deep, allows the latex to ooze out slowly and then hardens on the outside of the pod. Next day the hard latex is scraped off, before new incisions are made to obtain more poppy juice.

Purification, Chemical Composition and Properties of Morphine

Morphine was first isolated from opium in 1805 by a German pharmacist, Wilhelm Sertürner. He named it morphium—after Morpheus, the Greek god of dreams. The chemical structural formula of morphine was only determined in 1925. There are

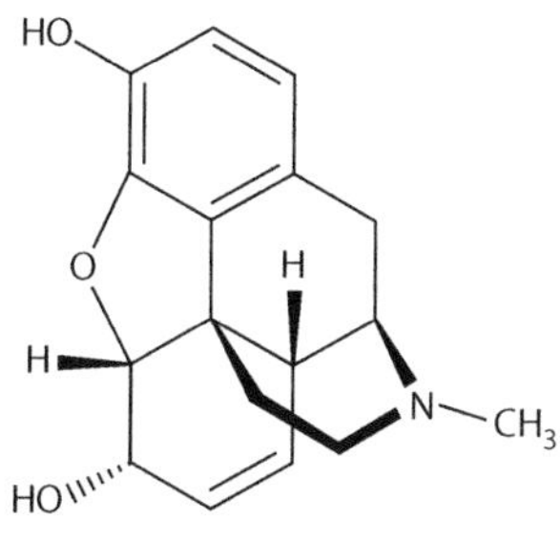

FIGURE 5.2 Chemical structure of morphine.

at least three ways of synthesizing morphine from starting materials such as coal tar and petroleum distillates, but the vast majority of morphine is derived from the opium poppy.

Morphine is purified from the opium resin in the following way: soaking the resin with diluted sulfuric acid, which releases the alkaloids into solution away from the plant and the resin without altering the alkaloid molecules. The alkaloids are now in the acidic solution and are then precipitated by either ammonium hydroxide or sodium carbonate. The last step separates morphine from other opium alkaloids. Today, morphine is isolated from opium in relatively large quantities; over 1,000 tons per year.

Chemical Composition and Properties of Codeine

Importantly, after sufficient amounts of pure morphine are retained for use as a pain killer, the remainder of the morphine produced for pharmaceutical use around the world is actually converted into codeine. Codeine is by far the most commonly used opioid in the world. As well as its use as an analgesic (painkiller), codeine is dissolved in a syrup-like liquid for use as a cough remedy.

Although codeine is found together with morphine, the concentration of the former in raw opium is much lower. Morphine can be converted into codeine using acetic anhydride, $(CH_3CO)_2O$. It is the hydroxyl group on the benzene ring of morphine which is acetylated.

Acetylation refers to the process of introducing an acetyl group into a compound through the substitution of an acetyl group for an active hydrogen atom. A reaction involving the replacement of the hydrogen atom of a hydroxyl group with an acetyl group, (CH_3CO), yields an ester, the acetate. Acetic (ethanoyl) anhydride is commonly used as an acetylating agent and will react readily with a hydroxyl group (see also the section on the emergence of aspirin, "Europe Solves a Headache!" in Chapter 2). The following reaction involving propanol and acetic (ethanoic) acid is an example.

$$C_3H_7OH + CH_3COOH = CH_3COOC_3H_7 + H_2O$$

The reaction would be much more vigorous with acetic anhydride

$$2C_3H_7OH + (CH_3CO)_2 = 2CH_3COOC_3H_7 + H_2O$$

Morphine can be converted into heroin by this acetylation reaction.

Chemical Composition and Properties of Heroin

A major search by scientists in Germany began in the late nineteenth century for a powerful alternative to opium and morphine and to find a form of the drug that could be taken orally instead of by injection.

Morphine can be converted into heroin (Figure 5.3). However, heroin is classified as an illegal recreational drug as it is highly addictive. In 1874, when morphine and

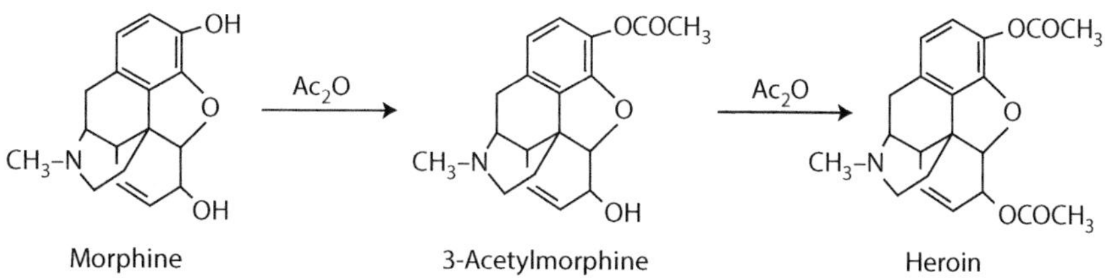

FIGURE 5.3 Acetylation of morphine to codeine and then to heroin using acetic anhydride.

acetic acid were boiled together, a chemical known as diacetylmorphine or morphine diacetate, also known as heroin, was produced. This is another example of an acetylation reaction. Both of the hydrogen atoms in the hydroxyl groups in morphine are replaced by acetyl groups, CH_3COO, to form heroin.

Heroin is approximately 1.5–2 times more potent than morphine weight-for-weight as it is able to cross the blood–brain barrier faster than morphine, subsequently increasing the reinforcing component of addiction. The drug is reconverted back to morphine in the body before it binds to brain tissue receptors.

Historically, Heinrich Dreser, who was in charge of drug development at Bayer, tested the new semi synthetic drug on animals, humans and even on himself. He pronounced heroin an effective treatment for a variety of respiratory ailments; especially bronchitis, asthma and tuberculosis. Since it was more potent than morphine, Bayer launched heroin in 1898 under its trademark as a new "wonder drug—the sedative for coughs" (Figure 5.4). Bayer was soon enthusiastically selling heroin in dozens of countries. Free samples were handed out to physicians. Sadly, the medical profession remained largely unaware of the potential risk of addiction for years. Eventually, doctors began to notice their patients were consuming inordinate quantities of heroin-based cough remedies. It transpired that heroin was not the miracle cure that some of its early promoters had supposed, and Bayer finally halted production in 1913.

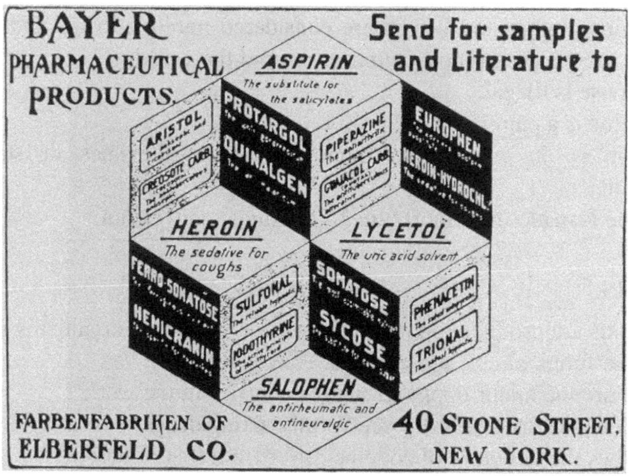

FIGURE 5.4 Bayer's heroin was sold legally at the turn of the twentieth century.

Medicinal Uses of Morphine

Morphine is used to relieve severe or agonizing pain and suffering and acts directly on the brain. It appears to mimic endorphins, natural substances produced by the brain that are responsible for reducing pain, which also cause sleepiness and feelings of pleasure in the body. Endorphins are released in response to pain, strenuous exercise or excitement. In clinical settings, morphine exerts its principal pharmacological effect on the brain. Its primary actions of therapeutic value are to reduce pain and make patients sleepy.

Morphine became a controlled substance in the United States under the Harrison Narcotics Tax Act of 1914. Possession without a prescription in the United States became a criminal offence. Morphine was the most commonly abused narcotic analgesic (pain killer) in the world until heroin was synthesized and came into use. Even today, morphine is the most sought-after prescription narcotic by heroin addicts when heroin is scarce.

Summary

Morphine is one of nature's great botanical miracles. It has powerful pain-killing effects and when used appropriately under medical supervision has been both a life saver and a giver of relief to millions of people throughout the world. Furthermore, the discovery of codeine as an antitussive compound (cough suppressant) has been a blessing to many people in combating coughs and related illnesses. Even though there are side effects, it appears that the benefits outweigh the risks. However, this cannot be said of heroin use, which has been a curse on societies and a destroyer of lives.

In the world of good drugs and bad drugs, this remains a tale of two cities:

- Most morphine is derived from the opium poppy.
- Morphine, opium, codeine and heroin are related chemicals known as alkaloids.
- Morphine, opium and heroin are considered narcotic drugs and are controlled substances in most countries around the world.
- Heroin use is illegal.
- Morphine is a pain killer.
- Codeine is an antitussive compound which alleviates or suppresses coughing.
- Codeine is made from morphine by chemical conversion.

QUESTIONS

1. Alkaloids are produced in nature by a large variety of organisms including bacteria, fungi, plants and animals. Why is this so?
2. Where are the major poppy-growing countries in the world?
3. Why does the drug industry seek orally acting drugs?
4. How does the structure of codeine differ from that of heroin?
5. What is thebaine? How is it different from morphine? Why is it an important alkaloid although not as biologically active as morphine?

REFERENCE

R. Cooper. 2013. Morphine & Heroin: The Yin & Yang of Narcotics. *ACS Chem Matters Dec*: 14–16.

SUGGESTED FURTHER READING

J. Mann. 2009. *Turn On and Tune In Pyschedelics, Narcotics and Euphoriants*. RSC Publishing.

C. Trocki. 1999. *Opium, Empire and Global Political Economy—A Study of the Asian Opium Trade 1750–1950*. Routledge, Taylor & Francis.

CANNABIS AND MARIJUANA

Abstract: An inebriant and modern social and medical enigma, cannabis—source of marijuana and hashish—is amongst the oldest of cultivated plants dating back to the beginnings of agriculture. Praised and maligned throughout history, cannabis continues to beguile us with its subtle magic of mystery and medicine.

Natural products chemistry

- Cannabinoids
- Terpenes.

Curriculum content

- Isomers
- Elimination reaction
- Condensation reaction
- Monoterpenes and the isoprene rule
- Cyclic terpenes.

CANNABIS

The use of fibrous material from cannabis goes back at least 10,000 years, and the consumption of cannabis as an intoxicant for at least 5,000 years. Charred seeds have been found in an ancient ritual brazier in Romania, while cannabis leaves and seeds have been discovered buried with a Chinese mummy.

Cannabis is a member of the hemp family of plants. Its fibers were turned into paper in China while the leaves were dried to form grass, which was then smoked for relaxation or used in teas for medicinal purposes.

In India, where the cannabis plant is known as *ganja*, a preparation of marijuana called *bhang* was made from the leaves and drunk with milk or water for its hallucinogenic effects. Extracts from the flowering heads were used in the treatment of anxiety as early as 1,400 BC.

Although the use of marijuana is controversial today, it does have some beneficial properties, which has led to its legalization for medicinal purposes in many countries. Specifically, cannabis is an established treatment for encouraging appetite in cancer patients. It is also utilized to help with nausea, weight gain, neuralgic pain and glaucoma.

INTRODUCING CANNABINOIDS

The cannabis genus has two economically important species: *Cannabis sativa* and *Cannabis indica*. These plants yield a sticky resin known as *hashish,* which contains a class of diverse compounds, some possessing hallucinogenic properties, known to science as cannabinoids. One particular cannabinoid is δ 1-tetrahydrocannabinol (THC) (Figure 5.5).

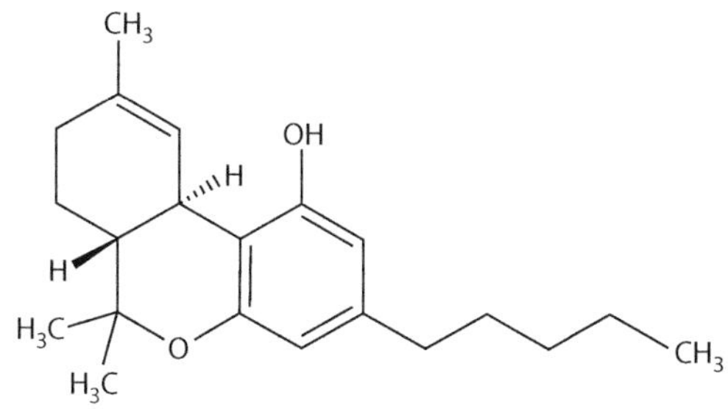

FIGURE 5.5 δ 1-Tetrahydrocannabinol (THC) is the main bioactive constituent of cannabis.

THC is the primary psychoactive compound in marijuana, and forms the natural biochemical defense of the cannabis plant against herbivores and disease. The discovery of THC was made by a team of researchers from the Hebrew University Pharmacy School, Israel, in 1964. Nowadays, synthetic THC is manufactured in California. It has been approved for limited medical use by the Federal Drug Administration in the United States. However, over 60 naturally occurring cannabinoids have been identified to date in cannabis. Several of these cannabinoids—in addition to naturally occurring terpenoids (oils) and flavonoids (phenols)—have also demonstrated therapeutic qualities. In fact, experiments have shown that the full range of psychoactive and medical effects of cannabis resin cannot be recreated by the singular use of pure synthetic cannabinoid drugs, such as THC. The indications are that other components of cannabis resin, such as terpenes, are either psychoactive themselves or are able to modulate the effect of cannabinoids when they are ingested together. Cannabinoids, such as THC, are chemically classified as terpenoids.

Terpenes

Terpenes are a huge and varied class of hydrocarbons that make up a majority of plant resins and saps. Essential oils (see the section on "European Lavender" in Chapter 6) composed primarily of terpenes, have a long history of medicinal use. A familiar terpene containing 40 carbons is β-carotene (see the section on "Saffron and Carotenoids" in Chapter 7), which is responsible for the orange color of carrots and is a source of "vitamin A." The low molecular weight terpenes, however, are very volatile and many can be recognized from their distinctive smell. Terpenes give pine trees and lemons their scent (pinene and limonene). Apart from use in fragrances (see the section on "European Lavender" in Chapter 6), terpenes may be employed as an organic solvent—oil of turpentine, a mixture of terpenes derived from pine trees, is an example.

Terpenes could find application as a replacement for crude oil as they are purely hydrocarbons. Many of the materials used today, whether as fuels, plastics, or deodorants, are derived from chemicals found in oil. Research is taking place to try and discover alternative, sustainable sources of basic substrates which can reduce or remove our dependency on oil. Substances in plants can be extracted and transformed into these industrial substrates. Sometimes these substrates are referred to as platform chemicals. They are simple and cheap, and are used by industry as starting materials from which more complex and valuable chemicals may be made.

Classification of Terpenes

Terpenes are polymers and are found mostly in plants (see also the section on "Saving the Pacific Yew Tree" in Chapter 2 and the sections on "Biblical Resins: Frankincense and Myrrh" and "A Plant from the East Indies, Camphor" in Chapter 6) although larger and more complex terpenes (e.g., squalene) do occur also in animals as steroids.

Terpenes are made up of multiples of isoprene molecules. Isoprene is an alkene having the molecular formula, $CH_2 \cdot C\ (CH_3) \cdot CH \cdot CH_2$, which has two carbon–carbon double bonds within a short five-carbon chain.

The simplest of all terpenes consists of two isoprene units linked together, and has the molecular formula $C_{10}H_{16}$.

Terpenes are classified by use of an empirical feature known as the isoprene rule. Molecules of terpenes, $(C_5H_8)_n$, such as squalene (presented below), are made up from linear multiples (n) of smaller isoprene molecules (C_5H_8), which each contains 5 carbon atoms (where n is an integer).

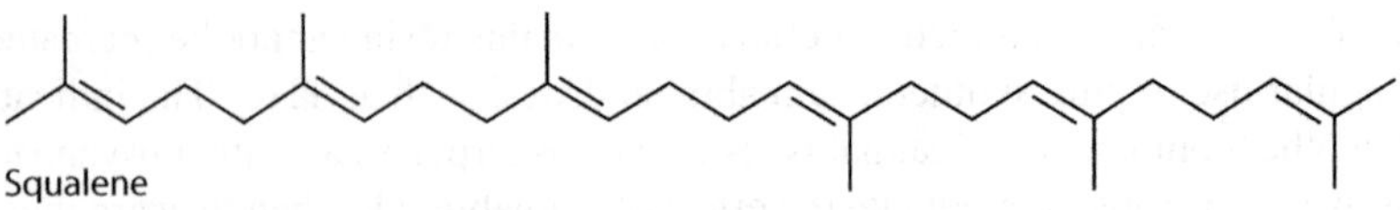

Squalene

If n = 2, the molecule is described as a monoterpene.
If n = 3, the molecule is described as a sesquiterpene.
If n = 4, the molecule is described as a diterpene, etc.

Terpenes can undergo natural biochemical modification through oxidation or complex rearrangement reactions to produce a variety of open chain and cyclic terpene compounds (see the sections on "Biblical Resins: Frankincense and Myrrh" and "A Plant from the East Indies, Camphor" in Chapter 6). In fact, a wide variety of cyclic terpenes is known. As terpenes, these molecules also consist of multiples of the building block, C_5H_8. However, cyclic terpenes do have fewer carbon–carbon double bonds than open chain or acyclic terpenes. An example of a cyclic monoterpene is the fragrant substance, limonene, (Figure 5.6), which may be obtained from the rind of lemons.

Terpenes and Elimination Reactions

In an elimination reaction, a small group of atoms (or a small molecule) break away from a larger molecule and are not replaced.

Dehydration is a common elimination reaction in nature. Elimination reactions, catalyzed by enzymes, lead to the formation of terpenes from carbohydrates and fatty acids. Although the details of the mechanisms of many of these reactions are

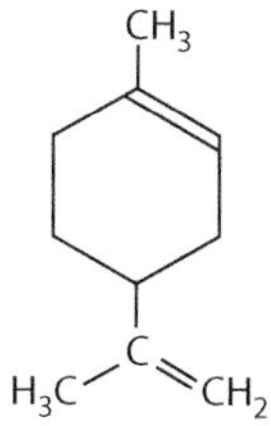

FIGURE 5.6 Limonene.

convoluted and little understood, the overall effect is the elimination of water, and hence the oxygen, from the carbohydrate or fatty acid. An illustration of dehydration in a simple elimination reaction, which can be performed in the laboratory, is the formation of the alkene, 2 butene, when the corresponding alcohol, 2 butanol, is heated in the presence of a strong acid such as sulfuric acid.

$$CH_3 \cdot CHOH \cdot CH_2 \cdot CH_3 = CH_3 \cdot CH \cdot CH \cdot CH_3 + H_2O$$

In the laboratory, elimination reactions are an effective means of introducing the functional group of the alkenes into molecules. Since alkenes are so reactive, they form a good platform from which to make many other organic chemicals. The elimination reaction is, therefore, a good starting point on a pathway toward the production of a whole array of organic compounds.

The Condensation Reaction

An elimination reaction is similar to a condensation reaction.

In a condensation reaction, the molecules of two compounds combine together with the loss of a molecule of water—that is why it is called a condensation reaction. There are many examples of condensation reactions in organic chemistry.

Condensation polymerization usually involves two different types of monomers. Each monomer has at least two functional groups—often at either end of the monomer. Each functional group reacts with a different functional group on a neighboring monomer to form a link which builds successively into a polymer chain. Examples of synthetic polymers formed from condensation reactions are polyesters and polyamides used in modern textiles and, of course, the polypeptides and proteins formed in the natural world, presented in the section on "Foods of the Fertile Crescent: Ancient Wheat" in Chapter 3.

QUESTIONS

1. Legalization of marijuana? For and against—discuss!
2. An illustration of dehydration is the formation of the alkene, 2 butene, when the corresponding alcohol, 2 butanol, is heated in the presence of a strong acid such as sulfuric acid.

$$CH_3 \cdot CHOH \cdot CH_2 \cdot CH_3 = CH_3 \cdot CH \cdot CH \cdot CH_3 + H_2O$$

Give examples of other elimination reactions in organic chemistry, which involve the loss of small molecules other than water.

3. A polyester may be formed from the reaction between molecules of a dicarboxylic acid, COOH·R·COOH, and the molecules of a diol, OH·R*·OH, where R and R* represent hydrocarbon chains.
   Draw the repeating unit of the polyester.
   Explain why this process is an example of a condensation reaction.
   Now, choose an amine and draw the repeating unit of the associated polypeptide.
4. The simplest of all terpenes consist of two isoprene units linked together and has the molecular formula $C_{10}H_{16}$. Provide a molecular structure for an isomer of this terpene.
5. Explain the term, addition polymerization.

SUGGESTED FURTHER READING

P. M. Richardson. 1986. *Flowering Plants—Magic in Bloom. Encyclopaedia of Psychoactive Drugs.* Chelsea House Publishers.

COCA AND COCAINE

Abstract: Over many centuries, coca was both a sacred herb to, and a staple part of, the diet of the indigenous peoples of the Andes (Colombia, Ecuador, Peru and Bolivia). Coca is the source of the useful but abused alkaloid, cocaine, which was first sold legally in the nineteenth century only to become listed as an illicit narcotic drug in the following century. And yes, with cocaine removed, coca is still a flavoring ingredient in Coca Cola®!

Chemistry of natural products

- The alkaloid, cocaine
- Organic compounds containing nitrogen in a carbon ring.

Curriculum content

- Alkaline nature of alkaloids
- Salt formation and acid–base extraction.

Coca and the Coca Plant

The coca bush has been a domesticated plant since early times in South America. Little was known of the narcotic properties of the leaf outside the Andes. People indigenous to South America chewed the leaves of the coca plant, *Erythroxylon coca*, as they are a source of vital nutrients in addition to numerous alkaloids, among them cocaine. Remains of coca leaves have been found interred with ancient Peruvian mummies, presumably because of belief in the value of coca in the afterlife. Chewing coca leaf was, and remains to the present day, a widespread practice in many indigenous communities in the high Andes. Although the stimulant and hunger-suppressant properties of coca had been known for many centuries, it was not until 1855 that the alkaloid responsible, namely cocaine, was first isolated (Figure 5.7). The first synthesis of cocaine was completed soon afterward, followed by elucidation of its structure in 1898.

Cocaine

Cocaine acts as a powerful stimulant, elevating mood sharply, but is quite addictive leading to a craving for more. A feeling of well-being, even euphoria, is quickly

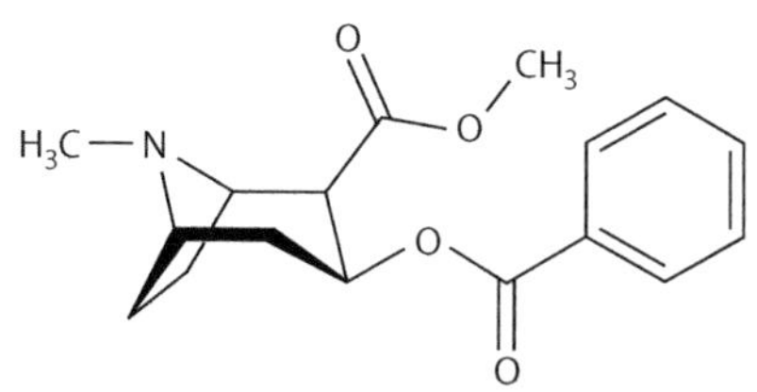

FIGURE 5.7 The alkaloid, cocaine.

followed by contrasting emotions—edginess, anxiety, paranoia and depression. Over time, abusers of cocaine experience a wide and disparate range of quite noxious physical effects that are too numerous to mention individually, but do include conditions associated with the cardiovascular toxicity of the drug, which involve heightened risk of cardiac arrest.

Cocaine was first traded as a legal drug in the nineteenth century, but as its damaging properties became better understood it was proscribed as an illicit narcotic in the 1980s. Now, production, distribution and sale of cocaine and cocaine products are restricted and illegal in most countries. The drug is regulated by the United Nations Convention against Illicit Traffic in Narcotic Drugs and Psychotropic Substances. Additionally, many countries have passed their own legislation. In the United States, the manufacture, importation, possession and distribution of cocaine is regulated by the 1970 Controlled Substances Act. As a consequence of suppression, it is the organized criminal cartels which dominate the supply of cocaine. Coca is grown and processed into cocaine in South America (particularly in Colombia, Bolivia and Peru) and, because of its high black-market price, is smuggled worldwide, particularly into developed countries, such as those in Europe and North America.

A synthetic means of producing cocaine would be highly desirable to illegal drug traffickers for obvious reasons, as it would eliminate their dependence on unreliable offshore sources and international smuggling. Fortunately, synthesis of economically significant quantities of cocaine in the laboratory is very difficult. The complex structure of the molecule gives rise to the unavoidable formation of many different enantiomers which are physiologically inactive, thereby severely limiting both the yield and purity of the product.

Chemical Properties of Cocaine

Alkaloids are a group of naturally occurring chemical compounds which contain nitrogen atoms within cyclic rings of carbon and hydrogen atoms. Alkaloids may also contain oxygen and sulfur. The nitrogen atoms in alkaloids behave in a similar fashion to the nitrogen atoms in amines (see the sections on "Coffee: Wake up and Smell the Aroma" in Chapter 4 and "Tobacco: a Profound Impact on the World" in this chapter) in that they are basic, although only weakly so. Curiously, most alkaloids have a bitter taste.

Although alkaloids are produced by a large variety of organisms, many of them are toxic to other organisms. In humans, many alkaloids exhibit pharmacological effects and have been used for centuries as medication, as recreational drugs and in tribal rituals. Various alkaloids act on a diversity of metabolic systems in humans, and several examples are covered in this book: the local anesthetic and stimulant, cocaine; the stimulants, caffeine and nicotine; the analgesic, morphine; the anticancer compound, vincristine; and the antimalarial drug, quinine.

Cocaine is an alkaloid derived from a natural building block called tropane. The alkaloids of tropane occur in certain families of plants; notably in the *Erythroxylaceae*, which includes coca, and also in the *Solanaceae*, which includes henbane, deadly nightshade, potato and tomato.

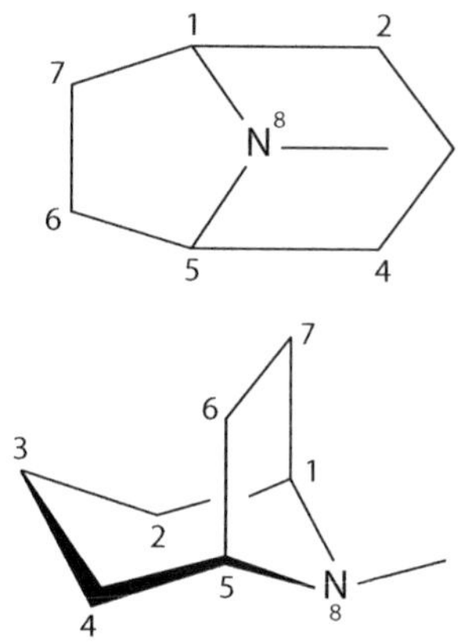

FIGURE 5.8 Molecular structure of tropane, shown both in planar form and in three dimensions, and also indicating the numbering of carbon atoms.

Tropane (Figure 5.8) is a bi-cyclic hydrocarbon with a single nitrogen atom bridging both a five-membered and a six-membered ring. In this respect, tropane is very similar to another heterocyclic saturated substance which contains a nitrogen atom in the ring, namely, piperidine, $C_6H_{10}N$. Piperidine is obtained from an extract of black pepper and is a secondary amine. Piperidine is used as a chemical building block for synthetic manufacture in the pharmaceutical industry.

As an alkaloid and a cyclic secondary amine, cocaine is a weakly basic compound, which can combine with stronger acids, such as inorganic acids, to form salts; the hydrochloride, sulfate and nitrate. The salts of cocaine are polar compounds. Therefore, they dissolve readily in a polar solvent such as water. In marked contrast, molecules of pure cocaine are covalent and are practically insoluble in water, yet are readily soluble in organic solvents.

Isolation of Cocaine

Acid–base extraction is a procedure using sequential, liquid–liquid extractions to purify acids or bases from mixtures by exploiting their chemical properties. The procedure works only for acids or bases with a large difference in aqueous solubility between their charged and their uncharged forms. Acid–base extraction can be performed as a routine process in the laboratory in order to isolate natural products such as specific alkaloids from crude extracts since alkaloids are weakly basic substances. Hence, the technique may be applied to isolate samples of pure cocaine.

This procedure is described extensively in the section entitled "Africa's Gift to the World" in Chapter 2, and in the section on "Maca from the High Andes in South America" in Chapter 4.

Legitimate Applications of Cocaine

It should be noted that according to the American College of Medical Toxicology, some limited and controlled medical applications of cocaine are permitted. It is used by some physicians to staunch strong nosebleeds in patients and as an anesthetic before minor nasal surgery. Again, due to its property as an anesthetic, dental surgeons may use it before oral procedures.

QUESTIONS

1. In looking at the structure of a molecule of cocaine, identify the functional groups and building blocks present.
2. Explain some of the chemical properties of cocaine by reference to the chemistry of a saturated heterocyclic compound, which contains nitrogen in the ring, namely piperidine.
3. Why is acid–base extraction helpful in purifying alkaloids?

SUGGESTED FURTHER READING

1. L. M. Harwood, C. J. Moody. 1989. *Experimental Organic Chemistry: Principles and Practice.* Wiley Blackwell.
2. H. Hobhouse. 2005. *Seeds of Change. Six Plants That Transformed Mankind.* Ch 6. Pages 291–363 Counterpoint.
3. J. Mann, J. Emsley, P. Ball, P. Page, J. P. Michael, H. Oakeley. 2009. *Turn On and Tune In Psychodelics, Narcotics and Euphoriants.* Royal Society of Chemistry, UK.
4. D. F. Rhoades. 1979. *Evolution of Plant Chemical Defense against Herbivores in Herbivores: Their Interaction with Secondary Plant Metabolites.* Academic Press.

TOBACCO: A PROFOUND IMPACT ON THE WORLD

Abstract: Another plant game-changer, tobacco, once introduced to the Old World, changed the habits of mankind forever, creating a powerful smoking habit through cigarettes, pipes and cigars, required slavery to meet ever-growing demand and, as modern science can attest, has led to nicotine addiction and to increased incidence of lung disease.

Natural products chemistry

- Nicotine
- Pyridine and pyrrole rings as building blocks.

Curriculum content

- Primary, secondary and tertiary amines
- Heterocyclic aromatic substances containing nitrogen or sulfur in the ring
- Amides.

Tobacco

Tobacco is an agricultural product processed from the leaves of plants in the genus *Nicotiana* of the Solanaceae family. The product manufactured from the leaf is used in cigars, cigarettes, snuff, pipe and chewing tobacco. The chief commercial species, *Nicotiana tabacum*, is believed to be native to tropical America. *Nicotiana rustica*, which is a mild-flavored and fast-burning species, was the tobacco originally grown in Virginia, although it is now cultivated chiefly in Turkey, India and Russia. The alkaloid, nicotine (Figure 5.9), is the most characteristic constituent of tobacco and is responsible for its addictive nature. The usage of tobacco today is practised by possibly up to one-third of the adult population worldwide.

The tobacco plant grows to a height of five feet and produces one harvest per year. The leaves are plucked and have medicinal value as an analgesic applied to wounds and snake and scorpion bites. The leaves were collected by natives of the New World, especially *Nicotiana tabacum*. The tobacco strain, *Nicotiana rustica*, spread through Central America northwards to North America and is believed to be the tobacco known in and around the Mississippi valley during the first century BC. Scientific evidence points to the fact that smoking and chewing tobacco were first practised on the American continent.

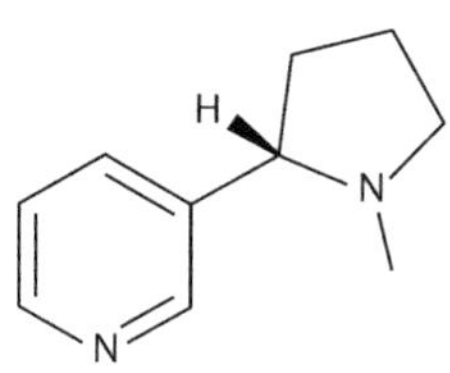

FIGURE 5.9 Nicotine, an alkaloid found in tobacco.

HISTORICAL NOTES

The Mayan Empire existed from the third to the ninth centuries AD. The Mayans lived in the Central American region, which we now know and recognize as the Yucatan peninsula of Mexico, Guatemala, El Salvador and Honduras. When the Spanish arrived, they found the Indian people to be ardent smokers using a herb unknown to the Western world. It was smoked recreationally as a pastime and also had significant religious and mythological implications. At the time of the Conquest, there was no cultivation, but today, it is most certainly a very important cash crop.

Columbus noted in his journal of 1492, that the San Salvadorian natives brought fruit, wooden spears and dried leaves (tobacco) of distinct fragrance. The local people smoked dried rolled leaves. The Spanish, and later, new arrivals to the New World, took up the smoking habit, which soon spread quickly among them.

The Spanish brought tobacco back to Europe although initially there was resistance to its use by the clergy. Nevertheless, by 1560, tobacco was being exported to Europe by Portuguese, Spanish, Dutch and English traders. In 1594, the historian, van Meteren, noted the use of tobacco in Holland: "…from diverse nations West Indies, Brazil and Peru, a dried herb called nicotania leaf (in the Indies known as tobacco) which was smoke-dried and used in a pipe lit with a live coal or candle."

Historically, the indigenous people were smoking a cigar form with large, rolled tubes of dried leaf. It appears that as tobacco use went northward, the preferred form of smoking was by pipe. The indigenous people of Mexico today are still predominantly cigar and cigarette smokers.

While tobacco had long been known in the Americas, it was not until the arrival of Europeans in North America, that tobacco became a widely abused drug and a very important item of trade. Its popularity initially stimulated the development of the economy of the southern states of the United States until it was superseded by cotton as an important cash crop. Following the American Civil War, changes in consumer demand and in the structure of the labor force allowed for the development of the cigarette and quickly led to the growth of tobacco companies.

Tobacco influenced the New to the Old World through commerce, simultaneously creating wealth for some and impacting adversely upon the health of others. Worldwide spread of the smoking of tobacco is a remarkable phenomenon in the cultural history of mankind. Less than 100 years after observing the New World Indians smoking tobacco, the practice became established almost universally.

Nicotine

Nicotine (Figure 5.9) was first isolated from the tobacco plant in 1828 by the physician, Wilhelm Heinrich Posselt, and the chemist, Karl Ludwig Reimann, in Germany, whilst its structure was elucidated in 1893 by Adolf Pinner and Richard Wolfenstein. The compound contains two nitrogen atoms, one in the larger pyridine ring and the second in a pyrrole ring. The systematic name for nicotine is 3-(1 methyl-2 pyrroldinyl) pyridine.

All alkaloids contain a nitrogen atom in the ring and are often referred to as tertiary amines. Thus, a short section on amines, in general, is warranted and discussed in detail, below.

Amines

Amines are ubiquitous in the natural world. Many important molecules are based on amines such as amino acids. Amines are utilized industrially as building blocks in the manufacture of dyes and pharmaceutical products.

Amines are compounds characterized by the presence of the nitrogen atom, a lone pair of electrons and three substituents. Amines are derivatives of ammonia in which one or more of the hydrogen atoms have been replaced with substitutes. Owing to the lone pair of electrons, amines are basis substances. The degree of basicity can be influenced by neighboring atoms, stereochemistry, and the degree of solubility in water of the cation produced.

Amine molecules can form hydrogen bonds. They are, therefore, soluble in water and have elevated melting and boiling points in relation to their mass.

Amines are classified according to the replacement of the hydrogen atoms bonded to the nitrogen atom in the following order:

- One substituent—primary amine
- Two substituents—secondary amine
- Three substituents—tertiary amine.

Amines are prepared as building blocks as the basis of a variety of industrial processes by alkylation with alcohols or by the reduction of nitriles by hydrogen.

In the case of cyclic amines, the nitrogen atom has been incorporated into a ring of carbon atoms effectively making the amine either like a secondary amine (in an aromatic ring) or a tertiary amine (in a nonaromatic ring).

Unsurprisingly, an amine group bonded to an aromatic group is known as an aromatic amine. The two functional groups influence one another: the aromatic structure decreases the alkalinity of the amine by the delocalization of electrons, while the amine group significantly increases the reactivity of the ring toward electrophilic agents for the same reason.

Heterocyclic Aromatic Compounds

Heterocyclic compounds, because they can contain one or more polyvalent atoms, besides carbon in a ring structure of any size, are exceedingly numerous: their chemistry, in fact, forms one of the largest subdivisions of organic chemistry.

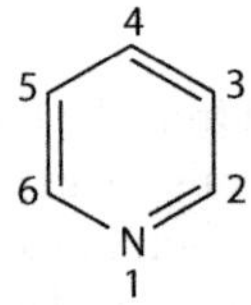

FIGURE 5.10 Molecular structure of pyridine indicating the numbering system of carbon atoms in the ring.

Pyridine

One of the most frequently occurring heterocyclic aromatic compounds is pyridine, (Figure 5.10), which has a structure very similar to that of benzene. However, the six-member ring contains a trivalent nitrogen atom.

Pyridine is a colorless liquid with an unpleasant, characteristic odor. It is completely miscible with water and most organic solvents, and is itself a very good solvent. As it is a stable, aromatic amine, pyridine is a moderate base in that it forms water-soluble salts, such as pyridine hydrochloride in the presence of hydrochloric acid.

Pyrrole

Pyrrole (Figure 5.11) is a colorless, volatile liquid which darkens on exposure to oxygen in the air. Pyrrole, C_4H_4N, is aromatic and has a five-membered ring.

As an aromatic amine, pyrrole is not very nucleophilic, being only weakly basic at the nitrogen atom. The pyrrole skeleton occurs commonly in molecules in the natural environment, for example, in chlorophyll and in vitamin B12—(see the section on "Our World of Green Plants: Human Survival" in Chapter 7 for more on chlorophyll).

Both pyridine and pyrrole are used extensively as chemical building blocks in the organic chemicals and pharmaceutical industries.

Thiophen

Another common heterocyclic compound is thiophen (Figure 5.12) which possesses a five-member ring containing one divalent sulfur atom (see the section on "Garlic and Pungent Smells" in Chapter 3 for more on sulfur-containing compounds in plants).

Owing to the delocalization of electrons over the ring structure, thiophen is remarkably stable. At this point, it will be worth referring to the properties of benzene described in the section entitled "Central America's Humble Potato" in

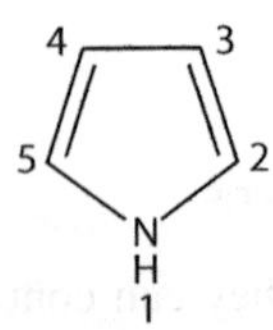

FIGURE 5.11 Molecular structure of pyrrole.

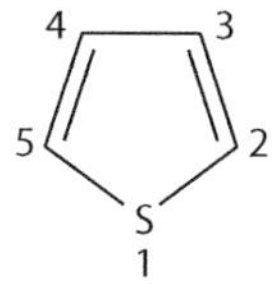

FIGURE 5.12 The molecular structure of thiophen.

Chapter 2. Pyridine, pyrrole and thiophen all show chemical behavior similar to that of benzene, and can undergo substitution reactions when hydrogen atoms on the ring are replaced by other atoms or groups.

Amides

An important chemical reaction of amines is the formation of amides with ketones or aldehydes, or more readily with acyl chlorides. The resulting compounds have a molecular structure in which a nitrogen atom is attached to a carbonyl group. These compounds are known as amides and have the general formula, $R1 \cdot CO \cdot NH_2$. In the case of amides substituted at the nitrogen atom, namely, N-substituted amides, the general formula is $R1 \cdot CO \cdot NR2R3$.

Owing to the higher electronegativity of the oxygen atom in comparison with that of the nitrogen atom, the carbonyl group is slightly dipolar with the carbon atom slightly positive, as electron density is displaced to the oxygen end of the covalent bond. As a consequence, carbonyl bonds react at the carbon atom with electron-rich entities, such as the nitrogen atom in the molecules of ammonia and amines, to produce amides.

Amines can be readily acylated by acyl chlorides to form amides and N-substituted amides.

$$CH_3 \cdot CO \cdot Cl + C_3H_7 \cdot NH_2 = CH_3 \cdot CO \cdot NH \cdot C_3H_7 + HCl$$

Ethanoyl chloride + propylamine = N-propylethanamide + hydrochloric acid

$$CH_3 \cdot CO \cdot Cl + C_6H_5 \cdot NH \cdot CH_3 = CH_3 \cdot CO \cdot NH \cdot C_6H_5 + HCl$$

Ethanoyl chloride + phenylamine = N-phenylethanamide + hydrochloric acid

Amides are common in the natural world. An example of a substance with an amide group is urea, NH_2CONH_2, which is one of the compounds excreted in urine as a result of the metabolic breakdown of proteins. Urea is used as an organic fertilizer in farming and, historically, in the tanning of animal hides to make leather. Amides are a major component of proteins and enzymes too—see the section entitled "Foods of the Fertile Crescent: Ancient Wheat" in Chapter 3.

An example of a synthetic amide of significant medical value for mild pain relief and fever reduction is known commercially as acetaminophen in the United States (but as paracetamol elsewhere) with a structure, $CH_3 \cdot CO \cdot NH \cdot C_6H_4 \cdot OH$, which can

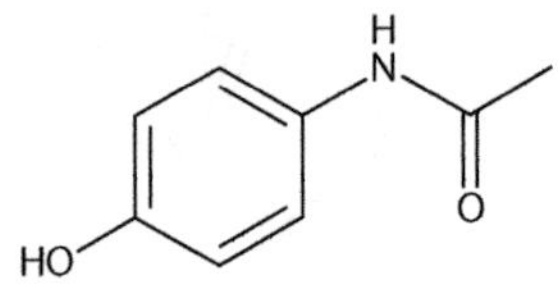

FIGURE 5.13 Acetaminophen or paracetamol.

be formed from a condensation reaction involving para-aminophenol and the acyl chloride, ethanoyl chloride (see Figure 5.13).

Properties of Nicotine

Nicotine is a hygroscopic, colorless oily liquid that is readily soluble in organic solvents such as alcohol and ether. It is miscible (see Glossary) with water between 60°C and 210°C. As a heterocyclic aromatic amine, nicotine forms salts and double salts with acids which are solid at room temperature and are water soluble.

The fused pyridine and pyrrole rings in nicotine are aromatic with electrons delocalized over the carbon–nitrogen skeleton, which helps to explain its weakly basic character and solubility in water.

Nicotine is optically active having two enantiomeric forms. The naturally occurring form of nicotine is levorotatory ((−)-nicotine) while the synthesized form, first prepared in 1904, is dextrorotatory ((+)-nicotine), and is less active physiologically. Evidently, stereochemistry plays an important part in the activity of (−)-nicotine.

On exposure to ultraviolet light or various oxidizing agents, nicotine is converted into nicotinic acid (vitamin B3) amongst other products.

Uses of Nicotine

Agriculture

Tobacco plants produce nicotine as a natural insecticide, and this can be concentrated for use as an artificial insecticide in order to improve the yield of crops grown in a monoculture. Neonicatinoids are synthetic analogues of the natural insecticide, nicotine. They are broad-spectrum, systemic, insecticides, which are applied as sprays to crops or as seed and soil treatments.

By contrast with agricultural methods supporting the growth of an economically important crop in a monoculture, organic farming relies on techniques, such as crop rotation, use of compost for soil fertility, and biological pest control. In the United States, the National Organic Standards Board (NOSB) defines the practice as follows:

> Organic agriculture is an ecological production management system that promotes and enhances biodiversity, biological cycles and soil biological activity. It is based on minimal use of off-farm inputs and uses management practices that restore, maintain and enhance ecological harmony.

Yields of crop per acre are lower, but organic farming is a sustainable practice. In recent decades, despite economic factors, the market for organic food and related products has grown rapidly. Demand has driven a corresponding increase in organically managed farmland, which has grown steadily in extent though still representing a relatively small proportion of farmland in cultivation worldwide.

Human Health and Pharmacy

There appears to be little compelling evidence that use of nicotine is related to a substantially increased risk of cancer. Nicotine is a stimulant producing a short-term increase in blood pressure and pulse rate, which could affect general health. Although nicotine is relatively safe for most individuals, it may have a negative effect on fetal development, and should be avoided during pregnancy.

It is true that a drop of pure nicotine can be deadly. The nicotine extracted from a pack of cigarettes and concentrated in one dose would be likely to be fatal if it were ingested all at once. Smokers are never exposed to pure nicotine, and do not absorb that much. However, the smoke from a burning cigarette contains compounds which have a devastating range of harmful health effects. The damage and health risks come from the smoke that is inhaled, rather than from the nicotine itself. This key fact is highlighted in an article in the American Council on Science and Health entitled "The Effects of Nicotine on Human Health." Smokers smoke for the nicotine "rush" yet, risk illness and early death from the effects of tobacco smoke. Ironically, a primary and current therapeutic use of nicotine arises in the treatment of nicotine dependence, in order to eliminate smoking, given the established damage that the practice does to both the health of the inhaler and to that of secondary inhalers. Controlled levels of nicotine are administered to patients through dermal patches, lozenges, electronic substitute cigarettes, or nasal sprays in an effort to wean them gradually away from dependence.

Nicotine has strong mood-altering effects, and can act on the brain as both a stimulant and a relaxant. Once in the bloodstream, nicotine will circulate around the body and reaches the brain within about 10 seconds. As is widely appreciated, nicotine is addictive. In the brain, nicotine stimulates the release of neurotransmitters, which relieve negative feelings, such as pain and anxiety, while pleasant sensations are enhanced. Nicotine does not produce intoxication, neither does it appear to impair judgment, motor skills or sociability. Nicotine intake reduces appetite, which, over time, causes weight loss. Because of these properties, nicotine is being considered as a therapeutic agent in small doses to treat such conditions as attention deficit disorder, Alzheimer's disease, Parkinson's disease, obesity, ulcerative colitis and inflammatory skin disorders.

QUESTIONS

1. Explain the nucleophilic substitution reactions of different aromatic heterocyclic compounds such as pyridine and pyrrole in terms of resonant structures and the Huckel theory.
2. Draw structural formulae for each of the following molecules:
    - Ethylisopropylamine
    - tert-Butylamine
    - 2-Aminopentane

- 1,6-Diaminohexane and
- N,3-Diethylaniline.

3. Draw diagrams showing how the molecules of anhydrous propylamine form hydrogen bonds to one another in the liquid phase. Then show schematically how the molecules in an aqueous solution of ethylamine would behave.
4. Highlight the methods and economics of organic farming, and compare and contrast them with those of standard agricultural practice which often occurs in monoculture.
5. Name and draw the structure of the molecule which has an oxygen atom instead of a nitrogen or sulfur atom within a five-membered ring.

REFERENCES

R. Powledge. 2004. Nicotine as therapy. *Public Library of Science Biology*, November 2(11).

S. Roberts. 2014. Review of literature. *News Medical.*

E. Whelan. 2014. The effects of nicotine on human health. *The American Council on Science and Health.*

SUGGESTED FURTHER READING

F. Robisek. 1978. *The Smoking Gods; Tobacco in Maya Art, History and Religion.* University of Oklahoma Press.

FIGURE 1.2 Yagua tribesmen in Iquitos, Peru can kill a monkey 30 m away with a blowgun. (With permission under terms of GNU Free Documentation License.)

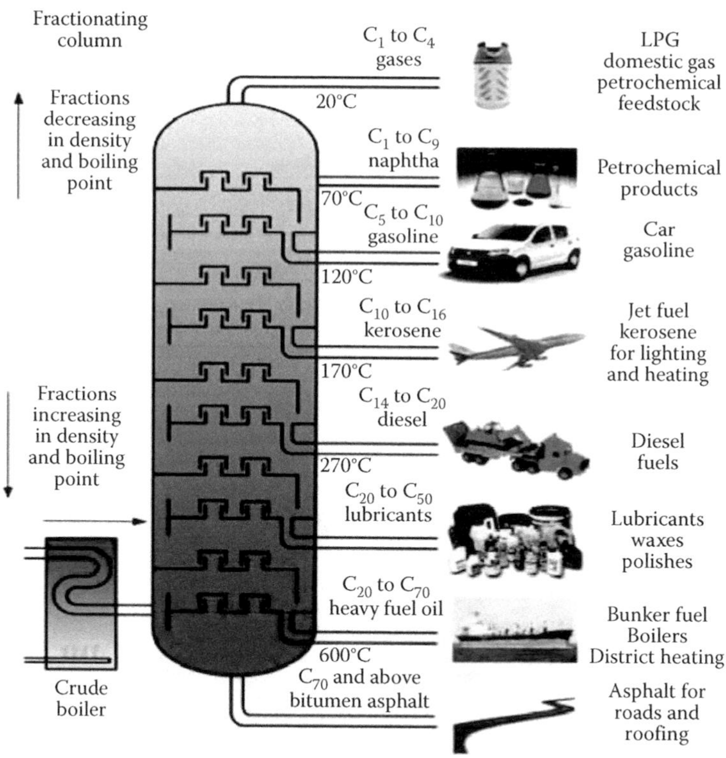

FIGURE 2.6 Diagram of an industrial fractionating column. (With permission under terms of GNU Free Documentation License.)

FIGURE 2.13 Parts of the world where malaria is endemic are shown in red on the map from the 2013 Global Malaria Mapper. (Courtesy of the World Health Organization (WHO), http://www.who.int/malaria/publications/world_malaria_report/global_malaria_mapper/en/)

FIGURE 2.15 *Artemisia annua* (annual wormwood). (Taken from R. Cooper and G. Nicola. 2014. *Natural Products Chemistry: Sources, Separations and Structures*. CRC Press, Taylor & Francis Group.)

FIGURE 2.18 Foxglove (*D. purpurea*). (Taken from R. Cooper and G. Nicola. 2014. *Natural Products Chemistry: Sources, Separations and Structures*. CRC Press, Taylor & Francis Group.)

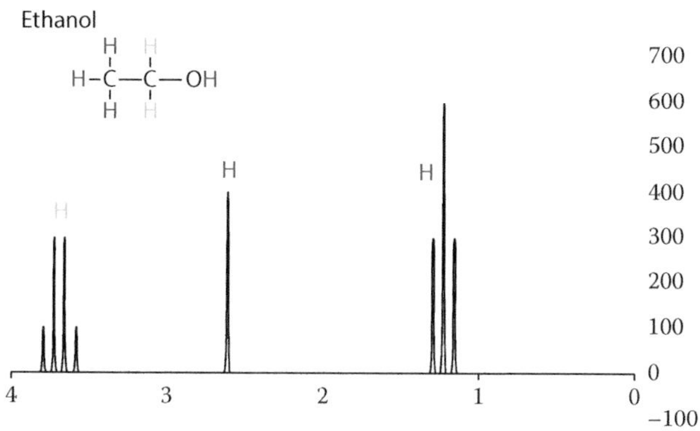

FIGURE 2.30 ^{1}H NMR spectrum of ethanol. (Taken from R. Cooper and G. Nicola. 2014. *Natural Products Chemistry: Sources, Separations and Structures*. CRC Press, Taylor & Francis Group.)

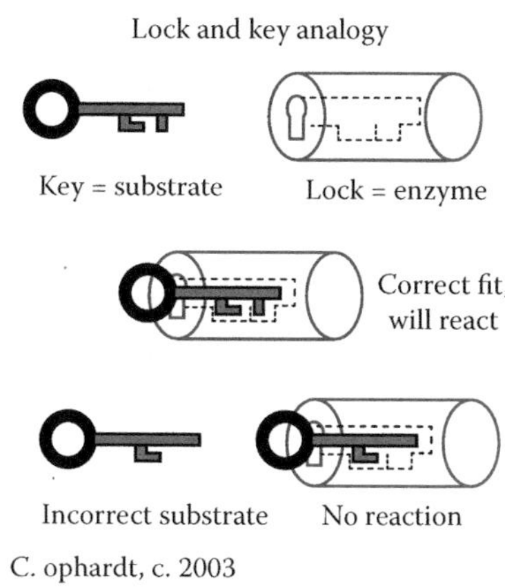

FIGURE 3.8 Schematic illustration of an active site in an enzyme—the lock and key model. (From http://www.elmhurst.edu/~chm/vchembook/571lockkey.html)

$$2NH_2COOH = NH_2CONHCOOH + H_2O$$

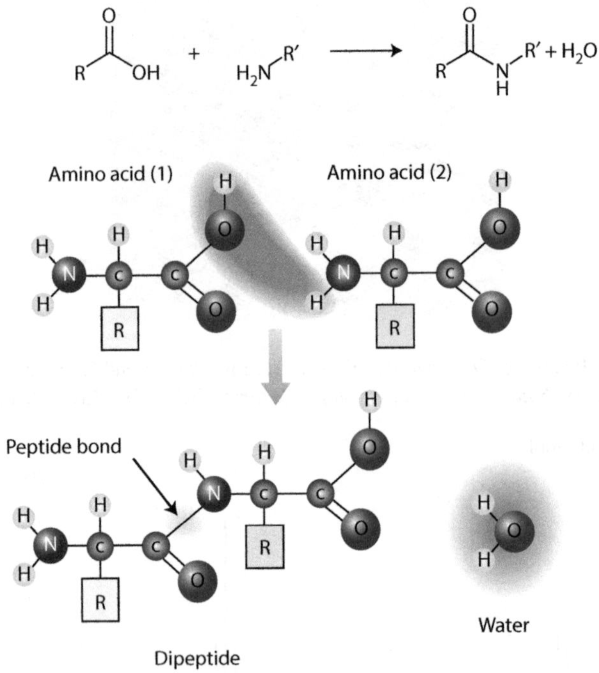

FIGURE 3.9 Formation of the peptide bond and formation of a dipeptide.

图 202 冬虫夏草 Cordyceps sinensis (Berk.) Sacc.
(示子座) (293)

FIGURE 3.19 Fruiting body of *C. sinensis*. (R. Cooper collection.)

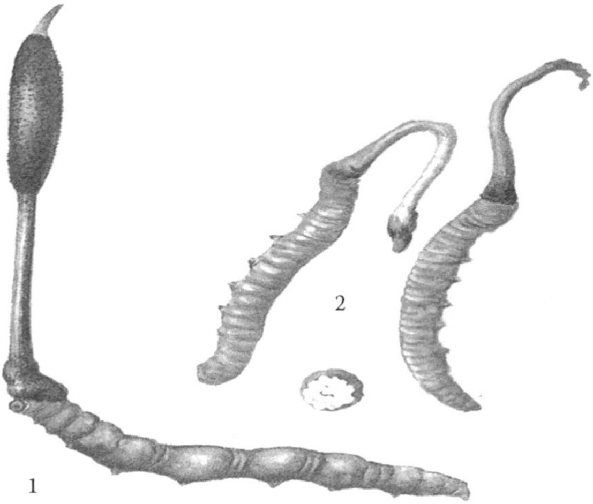

FIGURE 3.20 The fruiting body of *C. sinensis* (dark brown) protrudes from the Earth and grows from the caterpillar of the moth, *Paecilomyces hepiali* Chen (orange). (R. Cooper collection.)

FIGURE 3.22 Garlic bulbs, *A. sativum*. (With permission Steven Foster.)

FIGURE 4.4 *Coffea arabica*. (With permission from S. Foster.)

FIGURE 5.1 Poppy, *Papaver somniferum*, showing the petals and pods. (With permission from S. Foster.)

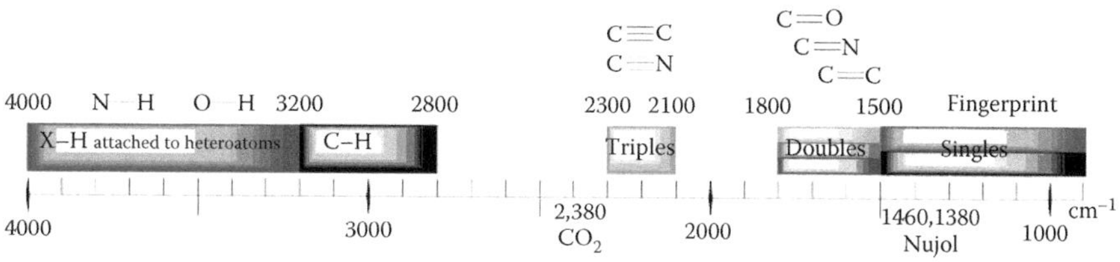

FIGURE 6.3 Infrared absorption frequencies.

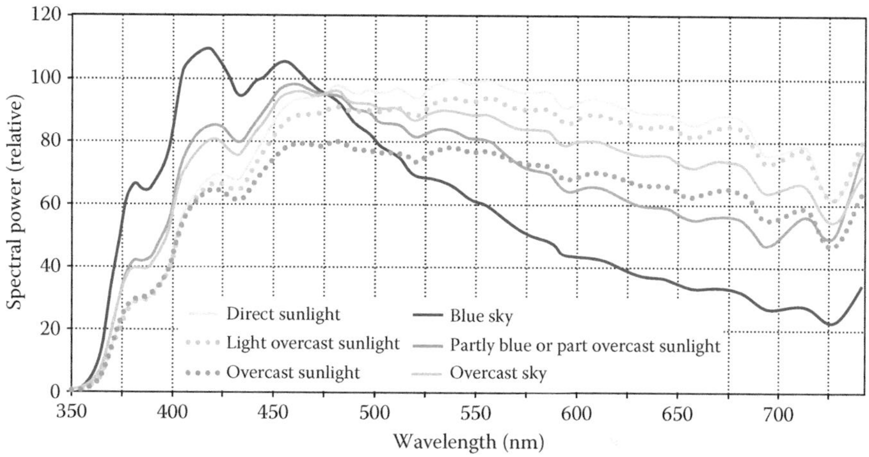

FIGURE 6.4 Spectrum of solar radiation.

FIGURE 6.13 Fields of lavender in full bloom. (With permission from S. Foster.)

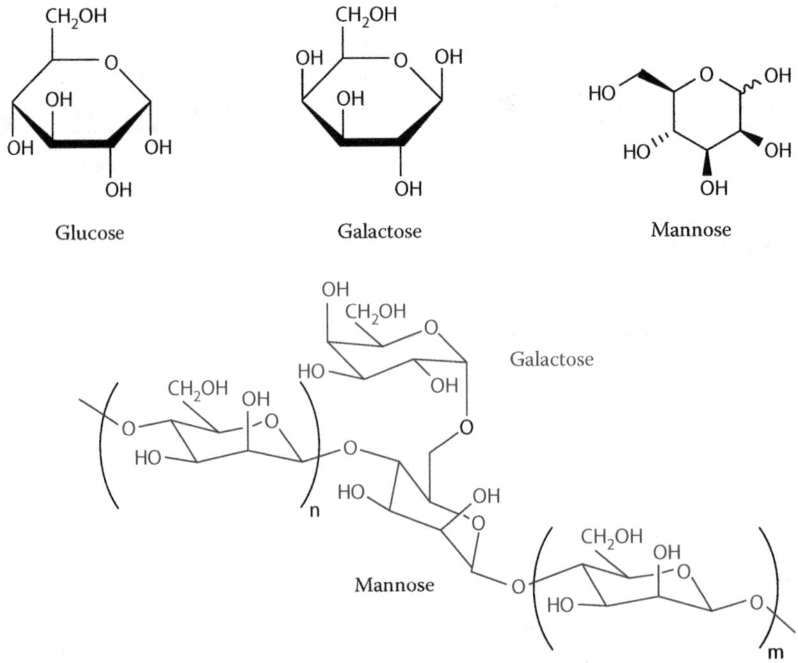

FIGURE 6.18 Three building blocks of sugars: glucose, galactose and mannose and an example of a polysaccharide structure, galactomannan. Note the sugars in blue constitute the linear backbone and in red galactose forms one of the side chains.

FIGURE 7.1 Natural pigments displayed in an autumnal scene at the National Arboretum, Gloucestershire, United Kingdom. (Courtesy of J.J. Deakin.)

FIGURE 7.4 Color in autumn leaves. (Courtesy of http://www.hdwallpapersos.com)

FIGURE 7.8 Attractive flowers of the Saffron crocus. (Courtesy of Dobies of Devon, UK; flower and seed merchants.)

FIGURE 7.17 Indigo from natural sources. (The historical collection of dyes at the Technical University of Dresden, Germany.)

FIGURE 7.23 Dyeing fabrics in Morocco.

FIGURE 7.27 The common lichen, *Xanthoria parietina*. (From http://www.freebigpictures.com.)

FIGURE 7.29 Purple pigmentation due to anthocyanins in the common pansy. (With permission from GNU Free Documentation License.)

FIGURE 7.30 Strawberries, raspberries, blackberries and blueberries. (Courtesy of http://www.wallpaperup.com)

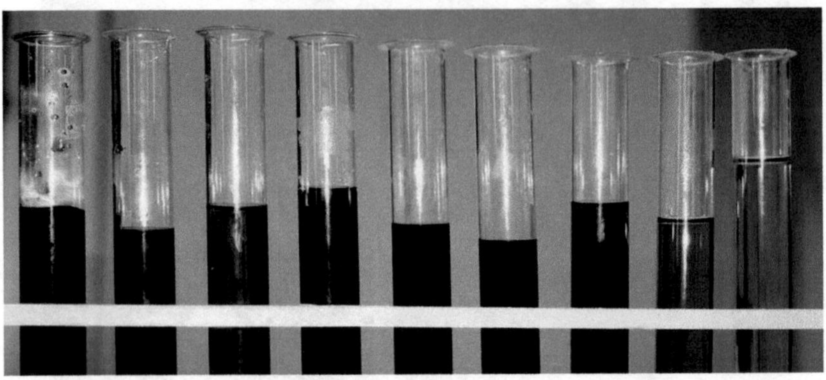

FIGURE 7.33 Anthocyanins from Red Cabbage change color reversibly dependent on the pH of the solution, from left to right 1, 3, 5, 7, 8, 9, 10, 11 and 13. (With permission from GNU Free Documentation License.)

6 Exotic Potions, Lotions and Oils

INTRODUCTION

It should come as no surprise that from the earliest times mankind sought chemical extracts from the huge natural resources of plants, which surrounded him. Plants are readily available, easily collected and harvested, and produce seeds for translocation and future investment. By trial and error, discoveries extended to exotic potions, lotions and oils, which made daily life more bearable and even enjoyable.

Appreciation of exotic fragrances and essential oils spread west from Asia. Trade in these materials formed a major part of commerce along the Silk Road and other routes.

The search for new exotic fragrances and essential oils continues apace today.

SUGGESTED FURTHER READING

K. Hughes. 2007. *The Incense Bible*. Haworth Press.

A PLANT FROM THE EAST INDIES: CAMPHOR

Abstract: Camphor is a volatile, waxy, white-to-translucent substance with a strong aromatic smell that is extracted from the wood of the camphor laurel, *Cinnamomum camphora*, a large evergreen tree of Sumatra and Borneo.

Natural products chemistry

- Terpenes
- Cyclic terpenoids.

Curriculum content

- Nucleophiles and nucleophilic addition reactions
- The carbonyl functional group
- Steam distillation
- Infrared spectroscopy and molecular structure determination
- "Greenhouse" effect, climate change and global warming.

The Classification and Structure of Camphor

Terpenes are a class of molecules that typically contain either 10 or up to 40 carbon atoms, formed from a five-carbon building block called isoprene (see also the section on "Saving the Pacific Yew Tree" in Chapter 2 and the section on "Cannabis and Marijuana" in Chapter 4).

Camphor (Figure 6.1) is related to terpene and is classified as a terpenoid with the chemical formula $C_{10}H_{16}O$. Its structure reveals the features of a cyclic alkane and a carbonyl functional group.

Terpenoids typically are found as components of plants and are regarded as essential oils, in that they have properties useful to man. Many terpenoids, for example, menthol and camphor, have medicinal value. A throat lozenge containing menthol will help clear blocked sinuses caused by the common cold. Wood varnish contains a terpene called pinene, which is abundant in pine trees. When pinene is exposed to air and sunlight, it oxidizes and polymerizes slowly to a mixture of terpenoids, which produces a fine hard finish for wood products.

Diverse Uses of Camphor

While camphor is poisonous if ingested in large doses, when utilized carefully in small quantities, it has many applications.

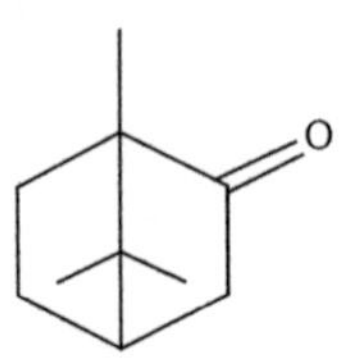

FIGURE 6.1 Structure of camphor.

The Food and Drug Administration (FDA) in the United States sets a low upper limit for the amount of camphor in consumer products. Medicinal use of camphor is discouraged by FDA except for skin-related applications, which contain only small amounts of camphor.

In medieval Europe, camphor was used as an ingredient in sweets. In Asian countries today, camphor is still used as a flavoring in confectionery and dessert dishes.

Camphor is readily absorbed through the skin, produces a soothing sensation of cooling, acts as a local anesthetic and is mildly antiseptic. It is this combination of properties which make camphor an effective ingredient of antiirritant and cooling gels for the skin. Camphor vapor is also an active ingredient in products, which can suppress coughing by relieving throat and bronchial irritation. Camphor is used also as a decongestant and as an essential oil in aromatherapy.

Modern applications of camphor also include its use as an insect and moth repellent to protect natural fabrics from damage, as an antimicrobial substance, hence its effectiveness as a preservative in embalming, and as a component in fireworks. In an enclosed space, solid camphor is in equilibrium with its vapor, due to its ability to sublime. In this way, a little camphor kept in tool chests will protect tools against rust as a thin film will sublime on a cold surface.

Extraction of Camphor by Steam Distillation

The presence of the single dipolar carbonyl group in an otherwise saturated cyclic molecule makes camphor only slightly soluble in water, a liquid composed of dipolar molecules. In fact, camphor may be regarded as immiscible with water. It is this physical property, which allows camphor to be extracted by a process known as steam distillation (Figure 6.2).

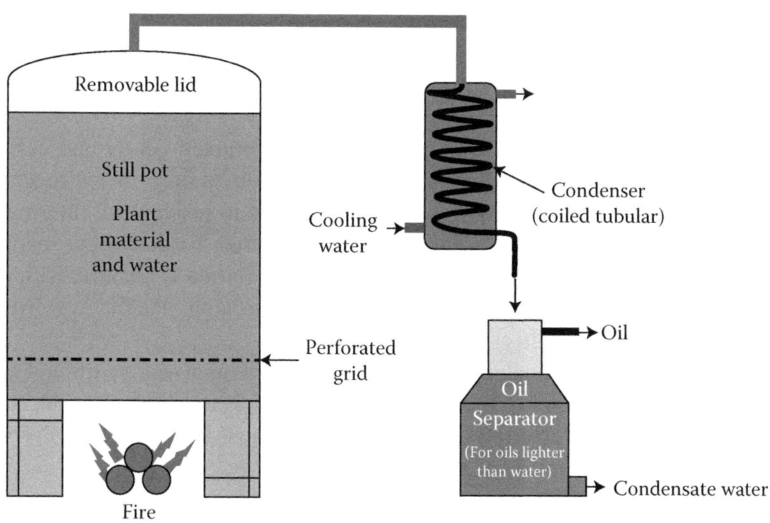

FIGURE 6.2 Illustration of the process of steam distillation. (Courtesy of Avraham Sand of AvAroma, Israel, http://www.AvAroma.com)

Camphor oil is isolated both in the laboratory and on an industrial scale by passing steam through the pulverized wood chippings of the camphor tree and condensing the vapors which arise. Camphor crystallizes from the oily part of the distillate and is then purified by sublimation (see Glossary).

Steam distillation is absolutely necessary because any attempt to distill camphor directly from the raw organic material would cause the camphor oil to decompose under the heat needed to bring it to the boil.

Owing to the fact that camphor oil and water are immiscible, each substance in the mixture vaporizes independently. It is well known that a liquid boils when the pressure of its vapor equals atmospheric pressure. Dalton's empirical law of partial pressures reminds us that the total pressure of a mixture of gases (or vapors) is equal to the sum of the partial pressures of the individual gases (or vapors). Effectively, the vapor pressures of each liquid in an immiscible mixture add together.

Hence, a mixture of two immiscible liquids will boil at a temperature lower than the normal boiling point of either component in the mixture (100°C and 209°C, respectively, for water and camphor). Because the vapor pressure of water is much greater than the vapor pressure of camphor oil, the mixture will boil at a temperature only just below the normal boiling point of water, causing the camphor oil to vaporize under moderate conditions, leaving the substance intact.

The Carbonyl Group and Nucleophilic Addition Reactions

The carbonyl group in a ketone (alkanone) and in an aldehyde (alkanal) is planar. It is also weakly dipolar and therefore, the double bond can undergo nucleophilic addition reactions.

A very useful step in organic synthesis is the addition of a carbon atom to a molecule.

$$CH_3 \cdot CO \cdot CH_3 + HCN = (CH_3)_2 \cdot C(OH) \cdot CN$$

An aqueous solution of hydrogen cyanide can be refluxed gently and carefully with a ketone in a fume cupboard, given the poisonous nature of hydrogen cyanide. The carbonyl bond is attacked by cyanide anions to produce an intermediate substance with a nitrile functional group, which can then be converted readily to other compounds. The nitrile compound produced in solution is racemic with equal concentrations of each of two enantiomers (see the section on "Pacific Yew Tree" in Chapter 2 for more on chirality).

In another example of a nucleophilic addition reaction involving a carbonyl group, an aldehyde can be reduced to a primary alcohol, or a ketone can be reduced to a secondary alcohol, by gently refluxing with a solution of lithium tetrahydroborate ($LiBH_4$) in a solution of water and methanol.

$$CH_3 \cdot CHO + 2H = CH_3 \cdot CH_2 \cdot OH$$

$$CH_3 \cdot CH_2 \cdot CO \cdot CH_3 + 2H = CH_3 \cdot CH_2 \cdot CH(OH) \cdot CH_3$$

The solution contains a rich source of hydrogen anions, which now attack the slightly positive carbon atom of the dipolar carbonyl bond.

Infrared Spectroscopy and the Determination of the Structure of Organic Molecules

Infrared spectroscopy exploits the infrared region of the spectrum of electromagnetic radiation where the frequency is lower than in the visible part of the spectrum and associated wavelengths are longer. When vapor samples of organic compounds are irradiated with infrared light, different bonds and functional groups in the molecule absorb energy in a narrow range of frequencies specific to them. The absorption of energy causes these bonds to vibrate with greater frequency. The drop in intensity in the infrared radiation at certain frequencies is detected by a spectrometer and is used as a "fingerprint" to identify the functional groups or bonds present providing improved understanding of the molecular structure of the compound under examination.

For example, in the spectrogram (Figure 6.3), the characteristic frequencies of absorption of the carbonyl bond, the nitrile bond, and the ethyne linkage are shown. In this way, the technique of infrared spectroscopy can assist greatly in the elucidation of the structure of an organic substance.

Absorption of Infrared Radiation: The Greenhouse Effect and Global Warming

The sun emits a wide spectrum of electromagnetic radiation (Figure 6.4), which includes the infrared (wavelength 2,500–700 nm), visible (wavelength 700–400 nm) and ultraviolet (wavelength 400–280 nm). One nanometer (nm) is one thousand billionth of a meter.

Much of the radiation from the Sun reaching the Earth (also called insolation) is absorbed in the higher reaches of the atmosphere by gases, free radicals and ions.

Some radiation is directly reflected back into Space from the Earth's surface by snowfields, by water in the oceans and by clouds, while some high-energy radiation is absorbed and reemitted at a lower energy in the infrared. Collectively, the radiation escaping the Earth into Space is known as the albedo (which is further discussed in the section on "Our World of Green Plants: Human Survival" in Chapter 7).

However, some of the direct and returned radiation is absorbed in the lowest, densest part of the atmosphere by certain atmospheric gases and is reemitted in

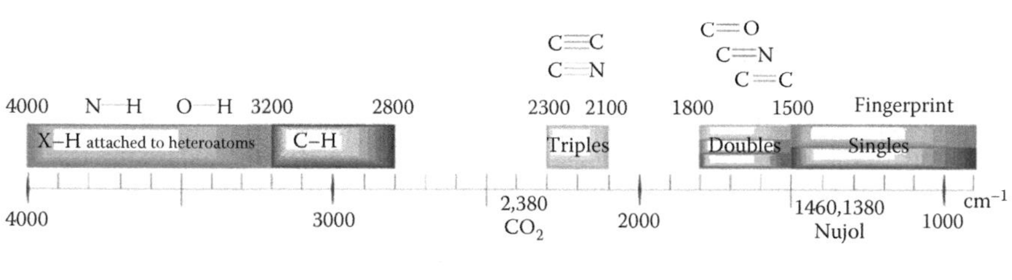

FIGURE 6.3 (See color insert.) Infrared absorption frequencies.

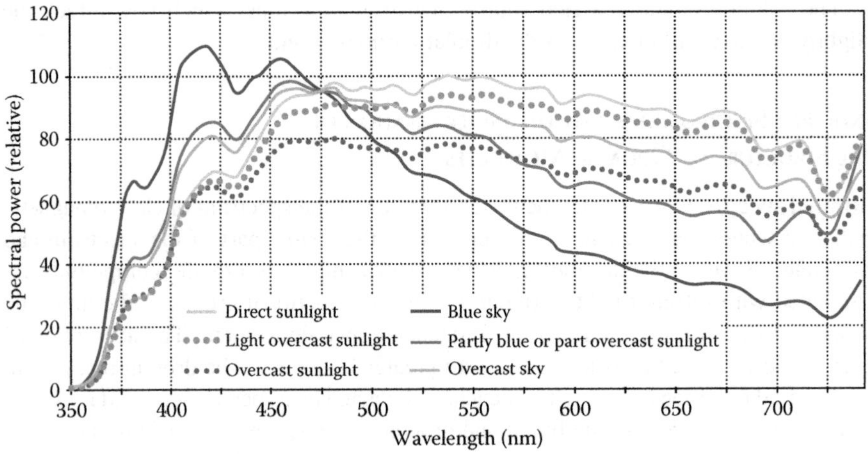

FIGURE 6.4 **(See color insert.)** Spectrum of solar radiation.

all directions including back to the Earth. Infrared radiation is strongly absorbed by water vapor, carbon dioxide, methane, nitric oxide and nitrous oxide (pollutants from vehicle exhausts) which are present in the atmosphere of the Earth. In turn, some of this absorbed energy (increased vibration in chemical bonds) is passed to other molecules, which are not active in the infrared through collision. The average translational or kinetic energy of molecules in the atmosphere increases, therefore, the temperature rises.

If the amount of these gases increases in the atmosphere disproportionately as a direct result of human activity, then the heating effect created makes a contribution to global warming—hence, the coining of the term—greenhouse gas.

Hydrocarbon bonds absorb particularly strongly in the infrared in comparison with the double bonds in carbon dioxide and the hydroxyl bonds in water vapor, so small, inadvertent emission of gaseous hydrocarbons is very significant. Yet, contributions to a rising level of hydrocarbons in the Earth's atmosphere can occur in the most innocent ways—even in agriculture. Cattle reared for beef or milk production release methane, as do the wet paddy fields where rice is grown. As more land is brought into cultivation in a populous world, so methane emissions rise. Ways of limiting global warming are discussed in the section on "Our World of Green Plants: Human Survival" in Chapter 7.

The mean temperature of the atmosphere, the oceans, and the surface of the Earth increases by almost imperceptible levels due to release of gaseous hydrocarbons (and especially methane). These gases are products of the decay of dead plants and organisms, and are released from the softening, partly frozen peat in the tundra and from sediments in the deep ocean. Methane is not soluble in water and enters the Earth's atmosphere, thereby accelerating the greenhouse effect.

In 2005, the first international treaty aimed at cutting greenhouse gas emissions came into effect. It is known as the "Kyoto Protocol" after the Japanese city in which discussions first began in 1997. The protocol seeks to control emissions of six greenhouse gases; methane, carbon dioxide, hydroflurocarbons, perfluorocarbons, nitrous oxide and sulfur hexafluoride.

Release of hydrocarbons into the atmosphere also causes damage to the ozone layer in the upper atmosphere. This was the subject of the "Montreal Protocol" in 1989 through which international agreement was reached to phase out, by the year 2000, use of hydroflurocarbons particularly as they are also greenhouse gases. Hydroflurocarbons had, until that time, been utilized as the propellant gas in aerosol cans and as the coolant material in refrigerators. It is encouraging to note that the Montreal Protocol has been reasonably successful, having been largely observed internationally by governments and business, who have sought practical alternatives to the continued use of hydrofluorocarbons.

Chemical pollution of the Earth's atmosphere by other means, through the release of the combustion products of organosulfur compounds for instance, is presented in the section on "Garlic and Pungent Smells" in Chapter 3.

ROYAL BOTANIC GARDENS: KEW, UNITED KINGDOM

The Royal Botanic Gardens at Kew in south west London is justifiably renowned for its extraordinary catalogue of plants and for its success in plant research and conservation. In response to continuing changes in plant habitat throughout the world brought about by climate change, the management of the Royal Botanic Gardens has invested heavily in the creation and development of a new facility, The Millennium Seed Bank, at Wakehurst Place in Sussex, England. This ambitious and visionary program collects seeds from all corners of the globe and stores them for posterity in carefully controlled conditions. In particular, global warming is gradually denying plants which live in delicate or marginal environments a stable and suitable habitat. Examples of such environments are

- The arctic and subarctic tundra
- High mountain chains, the Alps, Andes and Himalayas
- Microclimates in and around isolated valleys, or close to geophysical features such as waterfalls.

The aims of the Millennium Seed Bank are to conserve and preserve the seeds and spores of many, many thousands of species of plants, some of which are endangered, so that their characteristics and properties may be systematically studied both for their own sake and also to see whether chemical extracts from them would be of benefit to mankind.

QUESTIONS

1. Examine the molecular structure of camphor presented earlier in the chapter. There are three isomers of camphor, two of which rotate plane-polarized light. Can you identify each one?

2. When acetone is refluxed with an aqueous solution of hydrogen cyanide, the nitrile compound produced in solution is racemic with equal concentrations of each of two enantiomers. What are the conclusions that can be drawn from this fact about the mechanism of the attack of the nitrile anion, a nucleophile, on the carbonyl double bond? How would the presence of enantiomers be demonstrated in practice?
3. Explain the difference between infrared radiation and microwave radiation. Explain how the former is used to help identify the makeup of organic molecules, and why the latter is used to heat up substances, including food, and also as a carrier wave in telecommunications.
4. Explain what is meant by the term, greenhouse gas. Give a comprehensive account of the various ways in which the activities of humankind, directly and indirectly, are introducing more greenhouse gases into the atmosphere of the Earth. Describe potential solutions for reducing the level of emission of greenhouse gases including modified farming practice, and resort to alternative sources of power.
5. Explain how infrared spectroscopy can be used to monitor the concentration in parts per million of greenhouse gases in the Earth's atmosphere.
6. Chlorofluorocarbons (CFCs) were in widespread use during the twentieth century. Explain why the Montreal Protocol banned the industrial and commercial application of CFCs. Give three examples of their use and provide examples of alternative substances which are now in use.
7. Do you consider the national botanical gardens to be a valuable resource? Explain your reasoning.

SUGGESTED FURTHER READING

A. D. Cross, R. A. Jones. 1969. *An Introduction to Practical Infra-Red Spectroscopy*, 3rd edition. Springer Science.

K. Nakanishi, P. H. Solomon. 1977. *Infrared Absorption Spectroscopy*, 2nd edition. Emerson-Adams Press.

BIBLICAL RESINS: FRANKINCENSE AND MYRRH

Abstract: Frankincense and myrrh, essential oils from resins derived from shrubs of Arabia, were reputably gifts to the infant Jesus from the Three Wise Men.

Natural products chemistry

- Cyclic terpenoid acids
- Cyclic terpenols.

Curriculum content

- Essential oils
- Nomenclature of cyclic terpenes.

Sources and Uses of Frankincense and Myrrh

Both shrubs, frankincense, *Boswellia serrata*, and myrrh, *Commiphora myrrha*, belong to the family, *Burseraceae*. The scarcity of the resins of frankincense and myrrh meant that at one time they were more valuable than gold. This situation led to a continuing quest for the resins of frankincense and myrrh as exotic fragrances, and for burning as incense, and for use in embalming, respectively.

Incense refers to any aromatic plant material, which releases fragrant smoke when burned. Indeed, the term comes from Latin, *incendere*, meaning simply "to burn." Incense is still applied in a variety of ways in the modern world. It remains an integral part of the spiritual ceremonies of all the main religions. It is used as an air freshener, as an insect repellent, and in aromatherapy and meditation.

A BIBLICAL JOURNEY AND THE THREE GIFTS

The Magi (also referred to variously as the Three Wise Men, the Three Kings, the Three Astrologers, or Kings from the East) were a group of distinguished foreigners who were said to have visited Jesus after his birth, bearing gifts.

The Magi feature regularly in traditional accounts of the nativity celebrations of Christmas, and are an important part of the Christian tradition. Three gifts are explicitly identified in the Bible as gold, frankincense and myrrh: gold as a precious and attractive metal, frankincense as a perfume and myrrh as oil for anointing. The three gifts may also have a deeper meaning: gold, a symbol of kingship; frankincense, a perfumed incense representing spirituality; and myrrh, an embalming oil, a symbol of death and man's short time on Earth.

The scarcity of frankincense and myrrh might have suggested at one time they were more valuable than gold. Their scarcity led to a continuing quest for frankincense and myrrh as exotic fragrances and for incense.

Through spirituality, ancient civilizations have retained deep folkloric belief in frankincense: referred to as the male counterpart to myrrh's femininity. Many religions, worldwide, still use frankincense in their worship of the divine.

Frankincense has a long history of use in India as an Ayurvedic medicine. The herb is known as Gaja-bhaksha in Sanskrit and Sallaki Guggul in Ayurveda. Frankincense has antiinflammatory properties providing relief from the pain of rheumatism and arthritis. In manufacturing, Indian frankincense resin, oil and extracts are used in soaps, cosmetics, foods and beverages.

Historically, high-quality resin has been produced along the northern coast of Somalia and supplied to the Roman Catholic Church for use as incense.

Frankincense

The frankincense tree (Figure 6.5) is believed to originate from North East Africa and Southern Arabia where it is still found. The "weeping" of a gum from the frankincense tree yields a fragrant, aromatic resin (Figure 6.6). A volatile oil distilled from this resin is used in aromatherapy to counteract anxiety and relieve tension. The oil may be applied topically or inhaled.

Frankincense has a long history of use in India where it is highly regarded in Ayurvedic medicine. The herb is known as *Gaja-bhaksha* in Sanskrit and *Sallaki Guggul* in Ayurveda. Frankincense has antiinflammatory and analgesic properties, providing relief from the pain of rheumatism and arthritis. In manufacturing, Indian frankincense resin and extracts are used in soaps, cosmetics, foods and beverages.

FIGURE 6.5 Frankincense tree, *Boswellia serrata.* (http://www.shutterstock.com—stock number ID:51174910.)

FIGURE 6.6 Frankincense. (http://www.shutterstock.com—stock number ID:122765479.)

Olibanum is another name for frankincense, which arises from the milky resin exuded from incisions in the bark of several *Boswellia* species, including *Boswellia serrata*, *Boswellia carterii* and *Boswellia frereana*.

The exudate from *B. serrata* is most commonly used medicinally. Frankincense is tapped from the trees allowing the resin to seep out and harden. The hardened drops of resin are called "tears" because of their shape. Tapping is performed two to three times a year.

The principal, nonvolatile constituents of frankincense are α- and β-boswellic acid (see Figure 6.7) both of which have antiinflammatory properties. The boswellic acids constitute a group of carboxylic acids consisting of a hydrocarbon "skeleton" of a pentacyclic triterpene, and at least one other functional group. Both α-boswellic acid and β-boswellic acid have a molecular formula, $C_{30}H_{48}O_3$; they are isomers and differ only in their triterpene structure.

The boswellic acids are produced by the *Boswellia* family of plants and make up to 30% of the resin. It should be noted that while boswellic acids are major components of the resin they are not removed by steam distillation (see also the sections on "A Plant from the East Indies: Camphor" and "European Lavender" in this chapter). These acids are nonvolatile components and are too large to be carried over with the steam vapor during steam distillation.

The extracts of Boswellia are important in Ayurvedic medicine. There are some indications that derivatives of boswellic acid may have the properties of an anticancer agent. The long-term effects and side effects of taking frankincense have not yet been scientifically investigated. Nonetheless, several preliminary studies have been published, and one from researchers at the University of Leicester in the United Kingdom has been included in the references.

By contrast, the lighter terpenes from the resin are much more readily removed with the steam. Consequently, the essential oil collected as the condensate is richer

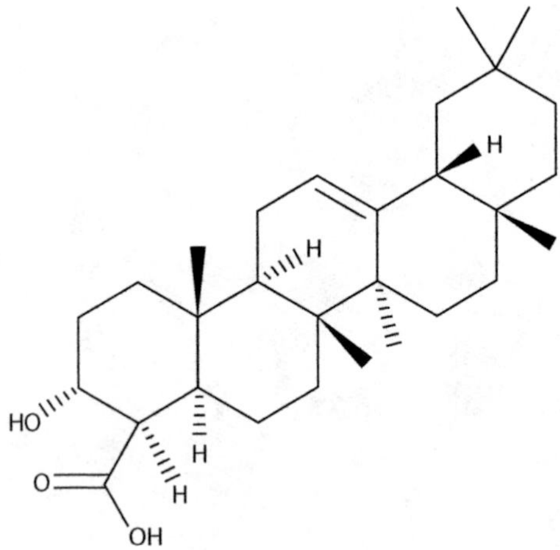

FIGURE 6.7 The structure of α-boswellic acid.

in the lighter molecules and consists of up to 75% of monoterpenes and sesquiterpenes. The most abundant substances are n-octylacetate, which is a straight-chain ester derived from octyl alcohol and acetic acid, having the molecular formula $CH_3(CH_2)_7COOCH_3$, and lesser amounts of 4-ethynyl-4-hydroxy-3, 5, 5-trimethyl-2-cyclohexen-1-one, discussed further (see Figure 6.10).

Myrrh

A thorny shrub about 3 m in height (Figure 6.8), myrrh produces pink and yellow flowers and beaked fruit. Its origin is believed to be Somalia, Ethiopia and Kenya.

The shrub exudes an aromatic, yellow resin from its stem which is also known as myrrh (Figure 6.9).

The term for the resin, "myrrh," derives from Aramaic (murr) and Arabic (mur) meaning "bitter."

The resin has astringent, antiseptic and antiinflammatory properties. These properties assist in treating mouth and throat infections and the common cold. In short, the resin may be used to treat mild inflammation of the oral and pharyngeal mucosa.

In commercial manufacturing, myrrh is applied as a fragrance and fixative in cosmetics. Importantly, myrrh continues to be used in embalming and as incense. Also, the essential oils from both myrrh and frankincense are used in aromatherapy.

In contrast to the essential oil of frankincense, 4-ethynyl-4-hydroxy-3,5,5-trimethyl-2-cyclohexen-1-one (Figure 6.10), is highest in abundance in the essential oil from myrrh.

Classification of Cyclic Terpenes

Terpenes are polymers and are found mostly in plants (see the sections on "Saving the Pacific Yew" in Chapter 2 and "A Plant from the East Indies: Camphor" in this

FIGURE 6.8 Myrrh tree, *Commiphora myrrha*.

FIGURE 6.9 Resin oozing from the bark of the Myrrh tree. (http://www.shutterstock.com—stock number ID:50096023.)

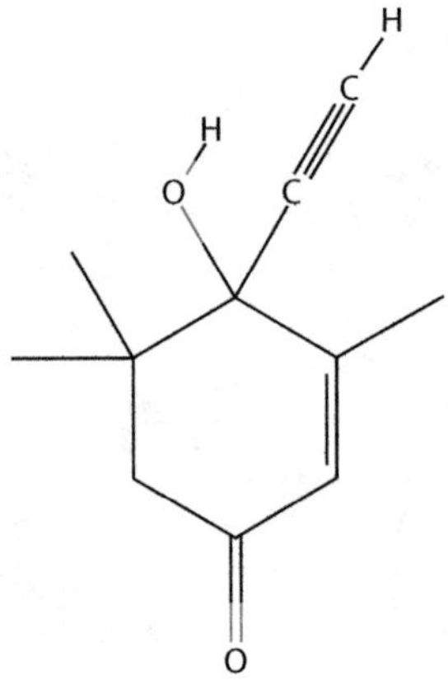

FIGURE 6.10 Major essential oil in the resin of myrrh.

chapter) although larger and more complex terpenes (e.g., squalene) do occur also in animals as steroids.

Terpenes consist of multiples of the isoprene molecule. Isoprene with the molecular formula, $CH_2{\cdot}C\ (CH_3){\cdot}CH{\cdot}CH_2$, is the name of a branched alkene, which has two carbon–carbon double bonds within a short, five-carbon chain.

Terpenes are classified by use of an empirical feature known as the isoprene rule. Molecules of terpenes $(C_5H_8)_n$, are made up from linear multiples (n) of smaller isoprene molecules (C_5H_8), which each contain five-carbon atoms (where n is an integer).

The naming (nomenclature) is a little confusing:

- Where n = 2, the molecule is described as a monoterpene
- Where n = 3, the molecule is described as a sesquiterpene
- Where n = 4, the molecule is described as a diterpene
- Where n = 5, the molecule is described as a triterpene, and so on.

The simplest open-chain terpene consists of two isoprene units linked together and has the molecular formula $C_{10}H_{16}$. Terpenes can undergo natural biochemical modification through oxidation or complex rearrangement reactions to produce a variety of open-chain and cyclic terpene compounds (see also the section on "A Plant from the East Indies: Camphor" in this chapter). In fact, a very wide variety of cyclic terpenes is known. As terpenes, these cyclic molecules are also made up from multiples of the building block, C_5H_8. However, cyclic terpenes do have fewer carbon–carbon double bonds than open-chain or acyclic terpenes. A straightforward example of a cyclic monoterpene is the fragrant substance, limonene (Figure 6.11), which may be obtained from the rind of lemons.

Returning to α-boswellic acid (shown in Figure 6.7 with the molecular formula, $C_{30}H_{48}O_3$), the structure contains five hydrocarbon rings and five terpene units. Hence, it is described as a pentacyclic triterpene.

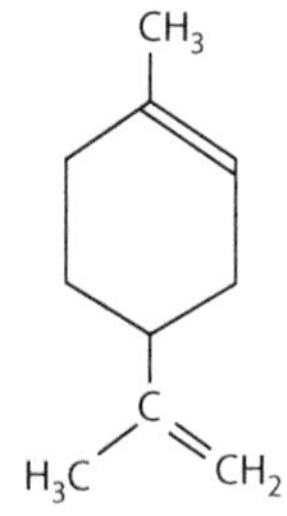

FIGURE 6.11 Structure of limonene.

Steric Hindrance

One of the most abundant chemicals present in the essential oils of frankincense and especially myrrh is 4-ethynyl-4-hydroxy-3,5,5-trimethyl-2-cyclohexen-1-one.

While the extract has an impressive systematic name, it is a cyclic terpenoid having a molecular formula of $C_{11}H_{14}O_2$. Unfortunately, in this case, there is not a more comfortable informal name. However, there is little steric hindrance (see Glossary) which means that the chemistry of the molecule reflects closely the chemistry of the functional groups present.

Steric hindrance can be exploited to change the pattern of the chemistry of a complex molecule by inhibiting an unwanted reaction involving a specific functional group (known as steric protection), or by leading to a preference for one course of stereochemical reaction over another (as in diastereoselectivity).

QUESTIONS

1. Why are the boswellic acids much less volatile than other chemical substances present in crude extracts of frankincense? Explain the value of this difference in the steam distillation of the essential oils of frankincense.
2. Give the systematic names for a simple open-chain terpene, $C_{10}H_{16}$, and a simple cyclic terpene, limonene.
3. Describe the chemical influences of the alkene, hydroxyl and carboxyl functional groups in boswellic acid.
4. 4-Ethynyl-4-hydroxy-3,5,5-trimethyl-2-cyclohexen-1-one is a chemical compound present, albeit in different amounts, in the essential oils of both frankincence and myrrh. From the systematic name of the compound, identify the functional groups which are present, and give a summary of the principal chemical properties of the molecule.

REFERENCES

N. Banno, T. Akihisa, K. Yasukawa et al. 2006. Anti-inflammatory activities of the triterpene acids from the resin of *Boswellia carteri*. *J. Ethnopharmacology* 107:249–253.

S. Hayashi, H. Amemori, H. Kameoka, M. Hanafusa, K. Furukawa. 1998. Comparison of volatile compounds from olibanum from various countries. *J. Essen. Oil Res.* 10:25–30.

S. Su, T. Wang, J. A. Duan et al. 2011. Anti-inflammatory and analgesic activity of different extracts of *Commiphora myrrha*. *J. Ethnopharmacology* 134:251–258.

SUGGESTED FURTHER READING

W. Albright, C. S. Mann. 1971. *Matthew. The Anchor Bible Series*. Doubleday & Company.

J. A. Duke. 2008. *Medicinal Plants of the Bible*. CRC Press.

A. Giesecke. 2014. *The Mythology of Plants: Botanical Lore from Ancient Greece and Rome*. Getty Publications.

N. Groom. 1981. *Frankincense and Myrrh: A Study of the Arabian Incense Trade*. Longman.

M. D. Herrera. 2011. *Holy Smoke: The Use of Incense in the Catholic Church*. Tixlini Scriptorium, San Luis Obispo.

G. A. Maloney 1997. *Gold, Frankincense, and Myrrh: An Introduction to Eastern Christian Spirituality*. Crossroad.

S. Mitchell. 2007. *A History of the Later Roman Empire*, AD 284–641: The Transformation of the Ancient World. Wiley-Blackwell.

B. E. van Wyk, M. Wink. 2004. *Medicinal Plants of the World*. Briza Publications.

EUROPEAN LAVENDER

Abstract: Extracts from the European lavender find use as a fragrant ingredient in toiletries, cosmetics and detergents; they are applied in medicine and are consumed as a flavoring agent in food and drink.

Natural products chemistry

- Essential oils
- Colloids and hydrosols
- Vegetable oils and fats
- Glycerol as a naturally occurring chemical building block.

Curriculum content

- Separation of essential oils by fractional distillation
- Hydrolysis and esters
- Saponification and soap
- Properties of detergents.

European Lavender

Lavenders (*Lavandula*) form a genus of 39 species of flowering plants belonging to the mint family, Lamiaceae. The genus includes annuals, herbaceous plants and small shrubs. The leaves are long and narrow in most species. The purplish blue color of the common lavender flower (Figure 6.12) is so well known that it has become eponymous.

FIGURE 6.12 Flowering lavender plants, *Lavandula angustifolia*. (With permission from S. Foster.)

HISTORICAL NOTE

The ancient Greeks knew lavender as the herb, nard, after the Syrian city of Naarda. Lavender was one of the herbs used in the temple to prepare a holy essence, and nard is mentioned in the Song of Solomon. The Greeks discovered early on that lavender would release a relaxing aroma when crushed or burned.

During Roman times, the flowers of lavender were sold for 100 denarii per pound, equivalent to a month's wages of a farm laborer. Lavender was often used in Roman baths to scent water and it was considered to be a skin restorative. The name lavender probably arises from the name of the plant in Latin, lavandarius, from lavanda (things to be washed) and from the verb, lavare (to wash). Lavender was introduced to England during the Roman conquest from AD 44 onward.

Distribution

Varieties of lavender flourish in dry, well-drained, sandy or gravelly soils, and can be found in many parts of Europe, North and East Africa, Arabia and India. The most widely cultivated species of lavender is *Lavandula angustifolia.* Although native to the western Mediterranean, the plant is commercially cultivated all over the world, including the British Isles, France, United States, Argentina and Japan (Figure 6.13).

FIGURE 6.13 **(See color insert.)** Fields of lavender in full bloom. (With permission from S. Foster.)

Value of Lavender

Food and Drink

Lavender flowers are occasionally blended with black tea, green tea or herbal tea, and add a fresh, relaxing scent and flavor. An infusion of flower heads added to a cup of boiling water makes a relaxing beverage.

Flowers can be candied to make "lavender sugar" and are sometimes used as cake decorations. Lavender is used to flavor baked goods and desserts.

Medicinal Applications

Lavender has had many uses in folk history, a number of which find application today. Infusions of lavender soothe and heal insect bites and burns. Bunches of lavender repel insects. When applied to the temples, lavender oil is believed to moderate headache. When applied on pillows, lavender seeds and flowers aid sleep and relaxation.

Oil of lavender is an antiseptic and, therefore, has antiinflammatory properties. It was used in hospitals during World War I to disinfect floors and walls.

In modern society, lavender is also used extensively in aromatherapy and massage therapy.

Aromatherapy is a form of alternative medicine in which healing effects are ascribed to the aromatic compounds in essential oils.

Cosmetics

The lavender plant, *L. angustifolia*, provides fragrant oil for balms, salves, perfumes and cosmetics. Lavandin, known as Dutch lavender, also yields oil with a higher level of terpenes, such as camphor, which add a sharp tone to the fragrance.

Soap and Detergent

The mildly antiseptic nature of lavender, and its pleasant, relaxing scent ensure that it continues to be widely used in quality soaps and toiletries.

Essential Oils: Hydrosol from Lavender

An essential oil is an aqueous mixture containing the volatile hydrophobic compounds of an individual plant, which are the "essence" of its distinctive aroma. Examples of popular essential oils are oil of clove, oil of eucalyptus, oil of peppermint, oil of spearmint, oil of cedar wood and, of course, oil of lavender. The distinctive scent of lavender is due to the presence of key monoterpenes (Figure 6.14), which are linalool, a terpene alcohol, and particularly linalyl acetate, a terpene ester.

The first step in the preparation of an essential oil from lavender involves subjecting the buds, spikes and flower tips of the plant to steam. The aqueous mixture produced is then fractionally distilled yielding a distillate, which is a mixture of essential oils and water. This mixture of essential oils is known as hydrosol.

Fractional distillation is based on the fact that different substances vaporize at different temperatures at a given atmospheric pressure, thus permitting separation of different parts or fractions of the mixture. Usually, fractional distillation is carried

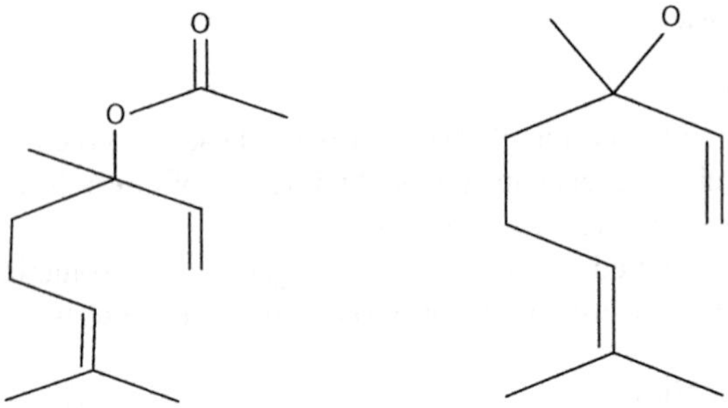

FIGURE 6.14 Linalyl acetate and linalool.

out when the boiling points of the fractions are in a relatively narrow range of temperature, say within 25°C. The vapor will contain higher concentrations of the more volatile components of the mixture, thus, the condensate collected will be enriched in the oils. Repeated fractional distillation would be necessary to separate fractions entirely. Typically, these principles are applied on an industrial scale in an oil refinery (Figure 2.6), where the many different fractions in crude oil are separated, and also in the cryoscopic fractional distillation of air, which can be liquefied at a very low temperature and high pressure. Cooled liquid nitrogen or oxygen, carbon dioxide and inert gases, such as neon and argon, are then kept under pressure in cylinders for further use.

Inevitably, the distillate collected from the first fractional distillation of an aqueous liquid mixture will contain various compounds, which vaporize and subsequently condense, at or below the temperature set for distillation. These compounds will include water, and the distillate will be a hydrosol.

Hydrosols and Colloids

Hydrosols are colloidal suspensions of essential oils in water. A colloidal suspension is formed when microscopic, insoluble particles of a solid or a liquid (the colloid) are dispersed throughout another substance (the continuous phase). The suspended particles are so fine that they do not settle, or only do so over a very long period of time. A colloidal suspension is entirely unlike a solution, where the solute dissolves in the solvent to form a single phase. Many colloidal suspensions are translucent because of the scattering of visible light by particles of the colloid. Milk illustrates this effect. It is a colloidal suspension of tiny, immiscible globules of butter fat dispersed in water.

Vegetable Oils and Fats

Vegetable oils and fats are widely distributed throughout the plant kingdom. They are complex mixtures of esters formed from medium- to high-molecular-weight carboxylic acids, and the chemical building block, glycerol, which is a trihydric alcohol. The distinction between oils and fats is simply a physical one arising from the

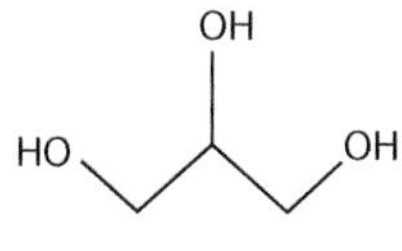

FIGURE 6.15 Glycerol, propane-1,2,3-triol.

difference in the scale of the molecular weight of the hydrocarbon chain; oils are normally liquid while fats are generally solid at ordinary temperatures and pressures.

Glycerol

Glycerol (Figure 6.15) is a colorless, odorless, viscous liquid resembling syrup in consistency. Owing to the polar nature of each of the three hydroxyl groups within a relatively small molecule, glycerol is very hygroscopic and is miscible with water in any proportion. Glycerol is used in the pharmaceutical industry for its softening and moisturizing properties and also as a lubricant. You can see from the figure that glycerol has the functions of both a primary alcohol and a secondary alcohol.

Triglycerides

In vegetable oils and fats, each of the three hydroxyl functional groups is now replaced by a long-chain carboxylic acid or fatty acid to make a triglyceride ester. Molecules of natural oils and fats typically involve three different fatty acids. The general formula of a naturally occurring triglyceride is shown in Figure 6.16.

Hydrolysis of Esters

Esters may be converted back to their constituent alcohols and carboxylic acids through the hydrolysis reaction. Specifically, a triester of glycerol would be heated in sodium hydroxide solution, a very strong base, producing the sodium salt of the weak acid and glycerol. Addition of a small amount of a dilute solution of an inorganic acid to the reaction products when cool will cause the sodium salt to precipitate as soap. The formation of soap in this way using an aqueous solution of an alkali has led to the hydrolysis of any ester being termed, saponification.

Soap

The soap produced by hydrolysis of long-chain esters is scented with the addition of an essential oil, which is often oil of lavender containing linalyl acetate. As an example, palm oil was used as the basis of the quality soap products made by the

$$\begin{array}{c} \mathrm{R.COO.CH_2} \\ | \\ \mathrm{R'.COO.CH} \\ | \\ \mathrm{R''.COO.CH_2} \end{array}$$

FIGURE 6.16 The general formula of a triglyceride.

Lever Brothers' company in the United Kingdom (now known as Unilever) which marketed "Sunlight" soap. In 1864, Caleb Johnson founded a soap company called B.J. Johnson Soap Co. in Milwaukee. In 1898, this company introduced a soap made of palm and olive oils marketed as "Palmolive." Later, the American company, Colgate-Palmolive acquired the brand.

Palm oil is a complex mixture of fatty acids, which include palmitic acid, $CH_3(CH_2)_{14}COOH$, and stearic acid, $CH_3(CH_2)_{16}COOH$. Oil palms are grown commercially in huge plantations in the tropics and especially in the East Indies. Palm oil is now used rather more extensively worldwide in processed food, for cooking oil and as a biodiesel fuel. However, the scale of production of palm oil has led to concern about negative environmental impacts, including:

- Deforestation
- Loss of habitat for endangered species such as the orangutan and Siberian tiger
- Potential for climate change through greater "greenhouse" gas emissions, though this is moderated to some degree through the absorption of carbon dioxide by the growing palms.

How Soap Works

Soap is a detergent. The term applies to any material which facilitates removal of dirt from the surface of something to be cleaned. In the modern world, there are many man-made materials other than soap, which are detergents.

Carboxylic acids and salts having alkyl chains longer than eight carbons exhibit unusual behavior in water due to the presence of both hydrophilic (carboxylic) and hydrophobic (alkyl) regions in the same molecule. Fatty acids made up of 10 or more carbon atoms are nearly insoluble in water, and because of their lower density, float on the surface of water. These fatty acids spread evenly over the water surface forming a very thin film, a monomolecular layer, in which the polar carboxyl groups are hydrogen bonded at the water interface and the hydrocarbon chains are aligned together away from the water. Substances that accumulate at water surfaces and lower the surface tension are called surfactants. By reducing the surface tension of water, surfactants allow water as the cleansing solvent to wet a variety of materials.

The alkali metal salts of fatty acids, as soaps, are more soluble in water than the acids themselves and are strong surfactants. For example, sodium stearate, $CH_3(CH_2)_{16}COONa$, is a soap with an anion attached to a long, nonpolar, alkyl stem. In small concentrations, soap, as a surfactant, will dissolve in water producing a random dispersion of solute ions. However, when the concentration is increased an interesting change occurs. The surfactant ions assemble in spherical aggregates called micelles. The hydrophobic chains of the ions form the center of the micelle, and the polar carboxylic ions extend into the surrounding water where they participate in hydrogen bonding.

Another problem relieved by soap relates to the natural presence of dissolved calcium and magnesium salts in the water supply (hard water). The metal cations are removed as an insoluble precipitate of the salts of carboxylic acids which is commonly recognized as scum.

In summary, the presence of soap in water facilitates the wetting of all parts of the object to be cleaned, removes water-insoluble dirt by incorporation in micelles and softens water by removing dissolved calcium and magnesium salts as scum.

QUESTIONS

1. What is meant by the term "essential oils"? Explain the physical difference between oils and fats.
2. Give an account of the principles lying behind the fractional distillation of a mixture of organic liquids. Show how fractional distillation may be done in the laboratory and is carried out in industrial applications. Also, explain how the process of cracking in an oil refinery is different to fractional distillation.
3. Account for the physical and chemical properties of glycerol which are dominated by the three hydroxyl functional groups present in the molecule.
4. Compare and contrast the properties of vegetable oils (esters of fatty acids), obtained directly from plants, with those of mineral oils (alkanes), obtained from deposits of crude petroleum.
5. Palm oil is used extensively worldwide in processed food, for cooking oil and as a biodiesel fuel. Discuss in a balanced way the arguments in favor of growing of crops for biofuel in relation to environmental concerns such as deforestation and the effect of net "greenhouse" gas emissions on climate change.
6. Explain how soap acts as a detergent in water.

SUGGESTED FURTHER READING

J. Bottero. 2002. *The Oldest Cuisine in the World: Cooking in Mesopotamia*. University of Chicago Press.

S. M. Cavitch. 1994. *The Natural Soap Book*. Storey Publishing.

S. Festing. 1982. *The Story of Lavender*. London Borough of Sutton, Libraries and Arts Services.

M. Lis-Balchin. 2002. *Lavender; The Genus Lavandula*. Taylor & Francis.

J. Rose. 2007. *Hydrosols and Aromatic Waters*. Institute of Aromatic & Herbal Study.

GLOBAL ALOE

Abstract: The aloe leaf has been adopted as the symbol of the Royal Society of Veterinary Medicine in the United Kingdom. Recorded in the Bible, used by the ancient Egyptians, fought over by Alexander the Great, this African plant was traded by the conquering Spanish army to South America, and is also found in traditional Chinese medicine. Aloe has many important medicinal properties which improve human and animal health.

Natural products chemistry

- Cosmetic oils
- Polysaccharides and glycosides
- Aloin and emodin.

Curriculum content

- Chromatography—thin layer, gas and gas liquid
- Ion-exchange properties.

Aloe vera

Aloe vera (Figure 6.17) is a short-stemmed succulent plant growing 50 cm to 1 m tall. The leaves are thick and fleshy, green to gray-green, with some varieties showing white flecks on their upper and lower stem surfaces.

The plant has been used in herbal medicine since the beginning of the first century AD. Carvings over 6,000 years old have been found in Egypt that contain images of the plant, referred to as the "plant of immortality." It was given as a burial gift

FIGURE 6.17 *Aloe vera.* (With permission from S. Foster.)

to deceased pharaohs. Furthermore, early records of the use of *Aloe vera* appear in various ancient texts including the Ebers Papyrus from the sixteenth century BC, in Dioscorides' "*De Materia Medica*" and Pliny the Elder's "*Natural History*" written in the middle of the first century AD.

Aloe vera and Skin Treatment

Extracts from *Aloe vera* are used widely in traditional herbal medicine in many countries. For example, *Aloe vera* in Ayurvedic medicine is used as a multipurpose skin treatment. Today, extracts from *Aloe vera* are widely used in the cosmetics and alternative medicine industries. It has rejuvenating, healing and soothing properties, and offers a moisturizing effect especially in cosmetic formulations where it can also provide relief from sunburn. The effect is due to the presence of polysaccharides in the extracts (also see the section on "An Asian Staple: Rice" in Chapter 3 for more on polysaccharides).

Aloe vera and Polysaccharides

Monosaccharide sugars may be linked through glycoside linkages to form polysaccharides. Glycosides contain a molecule of a sugar which is bound to another functional group via a glycoside bond. A good example of a glycoside is aloin, found in *Aloe vera*—described later in the section "Aloin."

Glycosides play numerous important roles in living organisms. Many plants store sugar in the form of inactive glycosides which can be broken down by hydrolysis promoted by an enzyme when the sugar is needed for use.

An important polysaccharide in the natural world is cellulose. Cellulose has the formula $(C_6H_{10}O_5)_n$ where n ranges from 500 to 5,000, depending on the source of the polymer. Over half of the total organic carbon in the Earth's biosphere is in cellulose found, for example, in cotton and plant fibers. The wood of bushes and trees contains roughly 50% cellulose. See the section on "An Asian Staple: Rice" in Chapter 3 for more on the glycoside link, polysaccharides and cellulose.

Galactomannans

The galactomannans form an important group of polysaccharides derived from aloe. These polymers are made up from the monosaccharide sugars, mannose and galactose, hence the name of the class (Figure 6.18). The spine of the polymer is essentially mannose with galactose attached as the side group. As a polysaccharide, galactomammans do not impart a sweet taste and do not readily dissolve in water in contrast to mono- and disaccharide sugars. It should be noted in discussing these monosaccharide sugars that glucose, mannose and galactose are isomers. Each has exactly the same molecular formula and molecular weight, but differs only in the configuration of their respective hydroxyl groups in the ring.

Galactomannans are also present in guar gum, which is a powder ground from the guar bean. The guar bean, *Cyamopsis tetragonoloba*, is a legume. Nodules on the roots of the bean fix nitrogen directly from the air due to the presence of bacteria

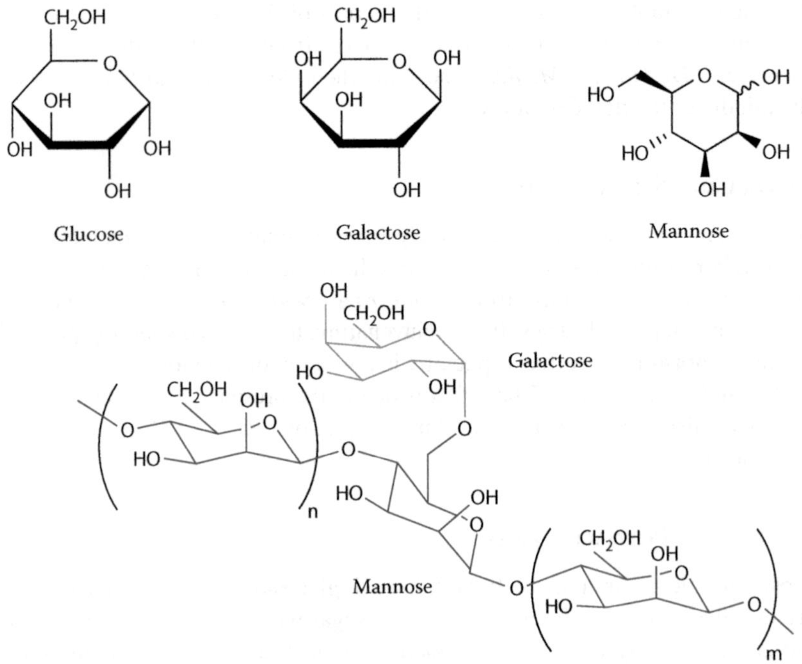

FIGURE 6.18 **(See color insert.)** Three building blocks of sugars: glucose, galactose and mannose and an example of a polysaccharide structure, galactomannan. Note the sugars in blue constitute the linear backbone and in red galactose forms one of the side chains.

and thereby increase the fertility of soil. Traditionally, the leaves and beans have been utilized not only as animal feed but also as a vegetable for human consumption. Historically, much of the world's supply of guar gum comes from India and Pakistan where the bean is cultivated.

When guar gum is added to water a thick gel is produced. The food processing industry takes full advantage of this property. Galactomannans are added to food products (such as ice cream, salad cream, sauces, tomato ketchup, soups and yogurt) to increase viscosity, and hence, improve texture.

Guar gum is also used for its medicinal properties as a mild laxative to help relieve chronic bowel conditions, such as colitis and Crohn's disease.

Hydraulic "Fracking"

A much more recent application of guar gum has sent its market price rocketing. The product has become important as an essential material in drilling for oil and natural gas through the process called hydraulic fracturing, known in common parlance as "fracking." In recent times, demand for guar gum has increased to such a degree that farmers formerly living in poverty in north western India and Pakistan have reaped a windfall, which has transformed their lives. Given that the price of guar gum has reached new heights, cultivation of guar beans is now spreading to

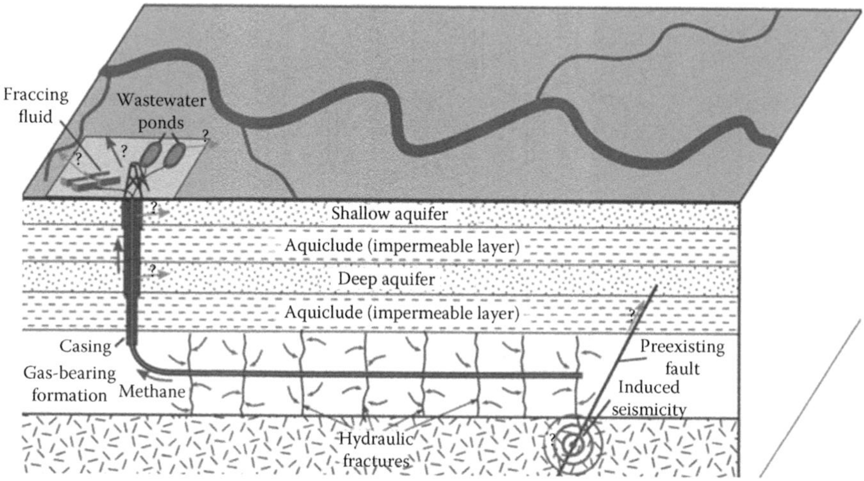

FIGURE 6.19 Schematic depiction of hydraulic fracturing for shale gas. (Courtesy of M. Nortor.)

other semiarid parts of the world including southern states of the United States such as Texas and Oklahoma. In hydraulic fracturing, a drill hole is initially bored into the Earth vertically. At a predetermined depth, the drill bit and the flexible drill line are then maneuvered into a horizontal position as drilling continues. A mixture of water and chemicals, including guar gum, is pumped down under enough pressure to release the trapped oil and natural gas by fracturing the soft shale rock in which it is contained. The pressure then forces the oil and gas through fissures to the surface where it is collected. The addition of guar gum to the water markedly increases the viscosity of the liquid making high-pressure pumping efficient, and the process of fracturing the rock more effective (Figure 6.19).

The amount of natural gas made available in this way may be great enough to significantly offset part of the energy demand of a developed industrial nation. Worldwide, it appears that reserves of natural gas held in shale rock could be considerable, which has reinvigorated debate about the ethics of exploitation of further reserves, given the concern about a greater degree of global warming through the accidental release of methane and other gaseous hydrocarbons, and the actual release of their combustion product, carbon dioxide.

Anthraquinones Found in *Aloe vera*

The anthraquinones are a common family of naturally occurring substances with yellow, orange and red pigmentation. The section on "Red Dyes: Henna, Dyer's Burgloss and Madder" in Chapter 7 provides much more detail on anthraquinones and on their application as dyes.

Aloin

Aloin (Figure 6.20) is an anthraquinone glycoside. It is extracted from *Aloe vera* as a mixture of two diastereomers, described as aloin A and aloin B. The reference

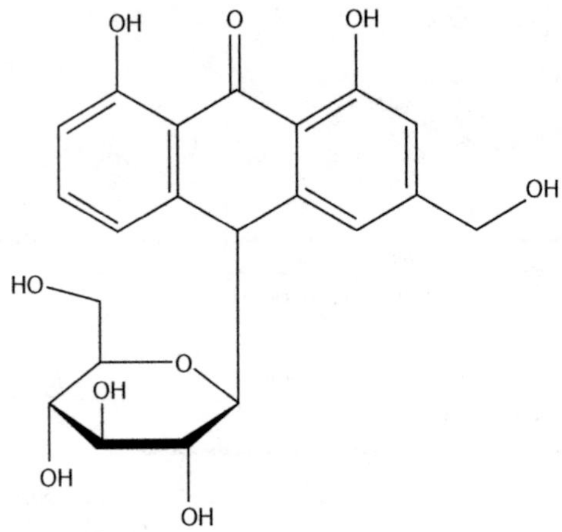

FIGURE 6.20 Structure of aloin.

to glycoside means that the polymer has an anthraquinone skeleton (the aglycone) which is then bound to a sugar molecule. In natural products chemistry, compounds possessing a sugar link are known as glycosides. Hydrolysis will generate the aglycone moiety and the free sugar moiety.

Aloin has been used as a traditional medicine since antiquity. It can be used in small amounts as a laxative. The juice of *Aloe vera* has been marketed to support digestive health although there is neither scientific evidence nor regulatory approval to support this claim. Aloin was a common ingredient sold in over-the-counter laxative products in the United States. However, in 2002, sales were prohibited by the FDA due to a lack of convincing safety data.

Aloin is a bitter, yellow-brown colored compound found in species of *Aloe*. Aloin is usually prepared by extraction from aloe latex that seeps out from just underneath the skin of aloe leaves. The latex is then dried and powdered to make the final product.

Emodin

The anthraquinone, emodin (Figure 6.21), is another purgative which can be obtained from the leaves of *Aloe vera* and, incidentally, also from the rhizome and stem of the rhubarb plant.

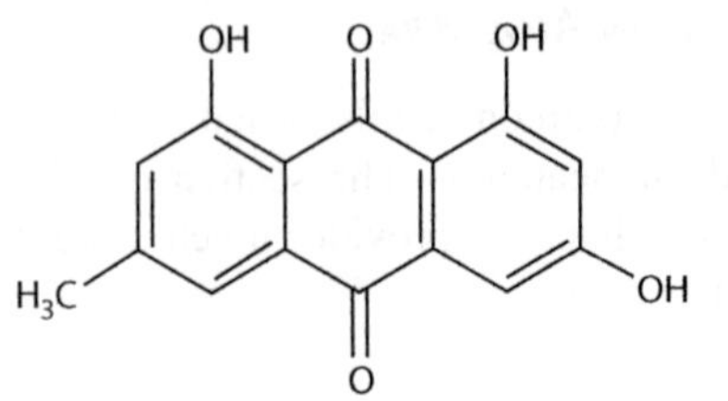

FIGURE 6.21 Structure of emodin.

The section titled "Red Dyes: Henna, Dyer's Bugloss, Madder" in Chapter 7 provides more on anthraquinones and on the use of emodin as a dyestuff.

Chromatography

Chromatography is a widely used method for separating and identifying samples of organic compounds, including polysaccharides and long-chain isomers, which are structurally similar. An important technical application of chromatography occurs in the pharmaceutical industry where the approach is used for analyzing the purity of drugs. Although chromatography is carried out in different formats (such as thin-layer, column, gas, high-pressure liquid, ion exchange), each technique depends upon common basic principles.

There are two essential elements:

- A stationary phase
- A mobile phase.

The mobile phase is fluid—a liquid or a gas—and contains the various dissolved substances to be separated. The mobile phase moves through or over the stationary phase in contact with its surface. In this way, the dissolved substances in the mobile phase are adsorbed at different rates on the surface of the stationary phase. As a consequence, the substances, which spend longer in the mobile phase travel further over the stationary phase. Separation is achieved.

Thin-Layer and Column Chromatography

The stationary phase is a layer of a solid, such as silica gel, positioned on a supporting structure of a smooth inert surface such as a plate of glass, hence the name, thin-layer chromatography. Alternatively, the solid material is packed into a vertical column, either loosely or tightly packed under pressure, hence the term, column chromatography.

The mobile phase is a solvent, such as ethanol, which passes over the stationary phase by upward capillary action, usually in the case of thin-layer chromatography, and downward under gravity in the case of column chromatography. Different solutes migrate at different speeds and can be separated. The identity of the solutes in the mobile phase is revealed by reference to tabulated records of

- How far they travel over the stationary phase compared to the solvent in thin-layer chromatography, or by
- How long it takes (known as the retention time) for the solute to move through the stationary phase in column chromatography.

Thin-layer chromatography is often used for separation of small quantities of material under analysis, and is a versatile technique for nonpolar compounds. The mobile phase, which is also known as the eluate, is commonly a single solvent, but could also be a mixture of miscible solvents.

Gas Chromatography and Gas–Liquid Chromatography

Gas chromatography and gas–liquid chromatography are techniques used for the separation of volatile, nonpolar compounds. If the stationary phase is a solid then the process is known as gas chromatography. When oil is used as the stationary phase it is referred to as gas–liquid chromatography. A chemically neutral carrier gas, nitrogen or helium, forms the mobile phase, which is passed through a heated, coated column or coil. A film of the stationary phase covers the interior of the long tube forming the coil, which is enclosed by an oven, so that the temperature can be regulated. Substances for separation dissolve at different rates in the stationary phase at a given temperature or remain in the gaseous state in the carrier, the mobile phase. This method has been particularly successful in the separation of volatile substances such as oils, esters of fatty acids and terpenes.

High-Pressure Liquid Chromatography

The advantages of high-pressure liquid chromatography arise from fast separation and use of small volumes of solvent.

Macroporous resins coated with suitable hydrocarbons are employed as the stationary phase with the size of the pores being selected beforehand to enhance the selectivity of separation. The mobile phase is often a mixture of miscible polar liquids, such as ethanol and water. Applications of the methodology include separation of long-chain, structurally related isomers.

Ion-Exchange Chromatography

Polar compounds are usually soluble in water. When the compound to be separated possesses an electrical charge as an ion or due to positive or negative functional groups within the molecule, ion-exchange chromatography is valuable. The stationary phase is a resin or silicate consisting of large polymeric molecules, often in the form of a cage structure. There are four main types of ion-exchange resin differing in the nature of the coating of the functional group, which make them active

- Strongly acidic (derivatives of sulfonic acid)
- Weakly acidic (carboxylic acid groups)
- Strongly basic (quaternary ammonium compounds, such as trimethyl ammonia)
- Weakly basic (primary, secondary or tertiary amines, such as polyethylamine).

Two approaches are possible; either the resin is positively charged, in which case the mobile phase or elute contains a negatively charged solute, or vice versa. In either case, the ionic strength of the mobile phase can be adjusted to alter the retention time.

QUESTIONS

1. Give a full account of the application of galactomannans in various industries including food processing, pharmaceuticals, cosmetics, explosives, paper manufacture and textiles.
2. Explain the role of galactomannans in the recovery of oil and natural gas by "fracking," and give a balanced account of the advantages and disadvantages of the exploitation of naturally occurring reserves of hydrocarbons by this process.
3. Describe fully the principles behind the important practical techniques of different forms of chromatography which is used so widely in the laboratory and in industry.

REFERENCES

K. Eshun, Q. He. 2004. *Aloe vera*: A valuable ingredient for the food, pharmaceutical and cosmetic industries—Review. *Critical Reviews in Food Science and Nutrition* 44(2):91–96.

L. E. Newton. 1979. In defence of the name, *Aloe vera. The Cactus and Succulent Journal of Great Britain* 41:29–30.

SUGGESTED FURTHER READING

R. Cooper, G. Nicola. 2014. *Natural Products Chemistry: Sources, Separations and Structures*. CRC Press, Taylor & Francis Group.

R. H. Davis. 1997. *Aloe Vera: A Scientific Approach*. Vantage Press.

Y. I. Park, S. K. Lee, Eds. 2006. *New Perspectives on Aloe*. Springer.

7 Colorful Chemistry
A Natural Palette of Plant Dyes and Pigments

INTRODUCTION

Have you ever wondered what makes the plant kingdom so colorful? Why are the leaves green and then turn to wonderful hues of yellow and red in the autumn? The explanation is due to the presence of natural pigments in the plants (Figure 7.1).

Chlorophyll is the green pigment of the plant kingdom. The compound gives green plants their characteristic color since it absorbs light in the red and blue parts of the visible spectrum of light, while reflecting green and yellow. Importantly, it provides the energy for photosynthesis of carbohydrates from carbon dioxide and water vapor in the atmosphere with the return of gaseous oxygen. In the autumn, the leaves lose their chlorophyll and as the green disappears the remaining natural red and yellow pigments become visible to the naked eye. The bright pigmentation of flowers, fruits and berries attract interest from animals, birds and insects, providing food as a reward in return for cross-pollination and the dispersal of seed.

FIGURE 7.1 **(See color insert.)** Natural pigments displayed in an autumnal scene at the National Arboretum, Gloucestershire, United Kingdom. (Courtesy of J.J. Deakin.)

Dyes and pigments owe their color to the selective absorption of certain wavelengths of visible light and the reflection of the remainder. Dyes are soluble in water and are usually applied in aqueous solution to a substrate to be colored, such as a textile. In contrast, pigments are insoluble and do not adhere to the surface of the substrate.

Historically, many dye plants were revered for possessing "magical properties" with the power to heal and to keep evil spirits at bay. Mystery and superstition surrounded the extraction of the essence. An example of this is provided by Woad, *Isatis tinctoria*, which yielded the purple-blue color, indigo, scarce in the natural world.

SELECTED READING

M. Pastoureau. 2001. *Blue: The History of a Color.* Princeton University Press.

K. Wells. 2013. Colour, health and wellbeing: The hidden qualities and properties of natural dyes. *Journal of the International Colour Association* 11:28–36.

OUR WORLD OF GREEN PLANTS: HUMAN SURVIVAL

Abstract: Directly and indirectly, members of the animal kingdom are wholly dependent upon plants. The plant kingdom is dominated by green plants converting carbon dioxide and water into a vast and diverse array of organic compounds providing vital resources in food, clothing and shelter, and releasing indispensable oxygen into the atmosphere for respiration.

Natural products chemistry

- Brief notes on the plant kingdom and green plants
- Lichens and symbiosis
- Chlorophyll
- Chlorin, a polymer of pyrrole, and a building block in nature.

Curriculum chemistry

- Photosynthesis
- Cage molecules—ligands and chelates (coordination compounds)
- Color and chromophores
- Light absorption and electron delocalization
- Ultraviolet absorption spectroscopy and the determination of molecular structure
- The exploitation of green plants and influence on climate change.

HISTORICAL NOTE ON THE PLANT KINGDOM

Robert Whittaker, a distinguished American ecologist, proposed in 1959 a macroscale classification of life in the natural world into five kingdoms:

- Animalia: mammals, birds and insects
- Plantae: green plants which include algae and seaweed
- Fungi: which obtain nutrition by breaking down organic compounds rather than by photosynthesis
- Protista: which contain disparate members that are unicellular or multicellular organisms without specialized tissues such as slime moulds
- Monera: which relate to unicellular organisms such as bacteria.

For decades, there has been debate about whether fungi may be regarded as plants or animals or something entirely separate from either, but Whittaker's classification appears to have achieved a degree of acceptance in the scientific community.

Members of the Plantae are remarkably diverse and may be broken down into two broad groups:

- Flowering plants and conifers, which reproduce from flowers and seed
- Algae, mosses and ferns, which reproduce from spores.

Lying outside the scope of this book, there are yet more subdivisions of these groups, which derive from consideration of the anatomical structure and function of plants.

Green Plants

Flowering Plants and Conifers

The importance of green plants to life as we know it cannot possibly be overstated. Green plants provide most of the world's oxygen, while flowering plants and conifers, numbering well in excess of 250,000 species, furnish much of the food (including fruit, berries, grains, nuts, leaves, stems, roots and vegetables) consumed by the members of the animal kingdom, the Animalia. In addition, humankind, in particular, derives benefit from green plants for clothing, shelter and fuel.

Algae

Algae (singular, alga), are a very large and diverse group ranging from unicellular genera, such as *Chlorella* and diatoms forming the phytoplankton of the oceans, to multicellular forms, such as the *giant kelp*, a large brown alga that may grow up to 50 m in length. Most are phototrophic (see Glossary) and lack many of the distinct cell and tissue types found in land plants, such as stomata, xylem and phloem. The largest and most complex marine algae are called seaweeds, while the most complex freshwater forms are the *Charophyta,* a division of algae that includes *Spirogyra* and the stoneworts.

Algae constitute a polyphyletic group as they do not include a common ancestor. Algae exhibit a wide range of reproductive strategies, from simple asexual cell division to complex forms of sexual reproduction.

Lichens

Lichens are communities of fungi and algae living together in symbiosis where each organism benefits from the presence of the other.

The fungus, a nongreen organism, provides structural support to the algal cells covering it, and provides nutrients, whereas the algae, green single-celled plants, provide carbohydrates through photosynthesis using its chlorophyll. Lichens can grow almost everywhere in the world from hot desert climates to snow-packed environments, but favor moist temperate zones where they can be found on rocks, roofs, walls and tree trunks. Many lichens have a green or orange tint.

The Vital Process of Photosynthesis

Joseph Priestley, the famous English chemist, showed, as early as in 1780, that green plants could "restore air, which has been injured by the burning of candles." He had discovered that green plants produce oxygen.

Later in 1794, Antoine Lavoisier discovered the concept of oxidation but was summarily executed during the troubled times after the French Revolution in 1789 on the grounds that "he was a monarchist sympathizer and the Republic of France had no need of scientists"!

The overall process of photosynthesis is now well understood and is summarized in the following chemical reaction, which takes place in the presence of sunlight and chlorophyll

$$6CO_2 + 6H_2O = C_6H_{12}O_6 + 6O_2$$

or more generally as

$$CO_2 + H_2O = (CH_2O) + O_2$$

Plants then use this glucose

- As a direct source of energy
- As the building block to drive various metabolic pathways including polymerization to form cellulose for growth
- By converting it into starch for the storage of energy in the roots
- To produce secondary metabolites for protection (see Glossary).

The vital process of photosynthesis is of paramount importance to animals and insects, which consume plant parts as food, and also use them as building materials for shelter.

Chlorophyll

Chlorophyll represents a family of very large and complex molecules located in the chloroplasts (see Glossary) of the leaves of plants. These compounds are responsible for the characteristic green color of plants. A representative structure of chlorophyll, known as chlorophyll A, is shown in Figure 7.2.

The key naturally occurring building block involved in the structure of chlorophyll is the chlorin ring which has the ability to coordinate with a central metallic ion. In the case of chlorophyll, the central doubly positive cation is that of magnesium. It is known that the magnesium ion plays a very important role in the metabolic pathways of plants.

Chlorin (Figure 7.3) is a particular example of a naturally occurring skeletal unit, known as a porphyrin, which is a polymer made up from essentially four pyrrole molecules (see Figure 5.11 in the section on "Tobacco: A Profound Impact on the World" in Chapter 5). Chlorin has a large cage-like structure. Incidentally, another example in nature of a complex ring similar to porphyrin is heme, which is a building block in the hemoglobin molecules present in the blood of animals. Four heme groups are

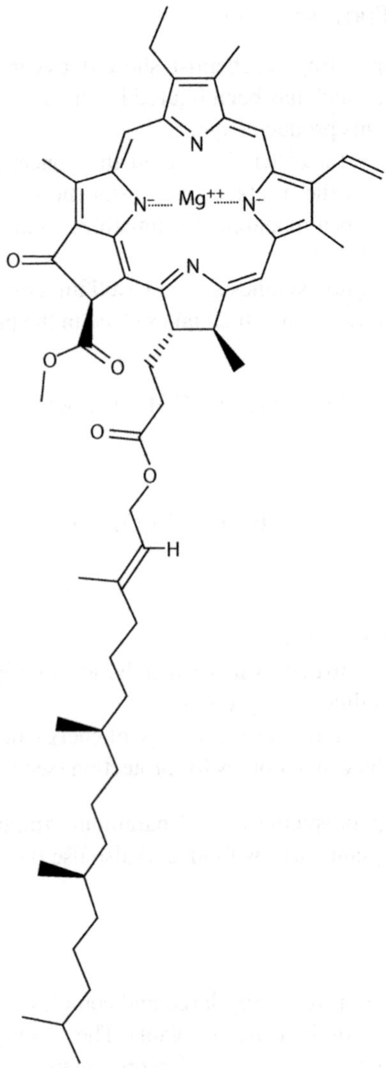

FIGURE 7.2 Molecular structure of chlorophyll A.

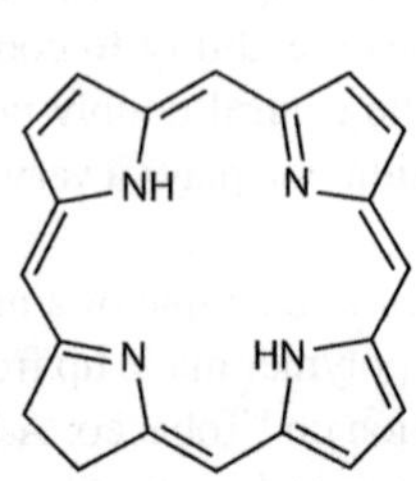

FIGURE 7.3 The cage-like structure of chlorin showing the linkages of four pyrrole groups in each corner.

bound to a protein called globin. In the case of heme, the metallic ion coordinated is that of iron. In the presence of oxygen, the ionic form of iron is red, whilst in anerobic conditions (see Glossary), namely within the veins rather than the arteries of the body, the ionic form of iron is blue—hence the color of blood. Thus, a coordination complex involving iron captures oxygen in the lungs of the animal in a reversible reaction and is instrumental in the chemical process of respiration in animals.

A molecule of chlorin has a cage-like structure with multiple, conjugated, unsaturated bonds: a network of alternating single and double bonds between carbon atoms, and a number of nitrogen atoms within the internal space.

Each nitrogen atom has a lone pair of electrons, which makes the internal central space (or interstice) of the chlorin molecule electron rich. This means that a small positive ion such as a magnesium ion can be incorporated by the chlorin molecule acting as a ligand; the complex ion, as a whole, being known as a chelate (see Glossary).

Ligands have at least one or more lone pairs of electrons available for bonding. Examples are water and ammonia (one each); ethane-1,2-diamine (two) and chlorin and heme (multiple). The lone pair of electrons belongs to the oxygen atom or the nitrogen atom in these molecules and is found in the outer valency shell. Since the spin of the two electrons is paired, the lone pair of electrons is not normally involved in covalent bonding, unless circumstances arise, as in coordination compounds, where the lone pair can be shared with a centrally located ion.

More on the role of coordination compounds and ligands in the organic chemistry of dyes can be found in the section on "Red Dyes from Henna, Dyer's Bugloss and Madder" in this chapter. Ligands are also involved in the action of neuroreceptors in the nervous systems of animals, and so reference is therefore made to the alkaloids presented in the section on "Maca from the High Andes in South America" in Chapter 4.

Chlorophyll and Color

Green plants absorb the blue and red parts of the visible spectrum of light due to two slightly different chlorophyll molecules, known simply as chlorophyll A and chlorophyll B.

Most of the visible light at 500–600 nm in wavelength is reflected, and this is green—hence the color of plants. The reflection of green light is strong enough to mask the effects of weaker absorption of light by carotene (see the section on "Saffron and Carotenoids: Yellow and Orange Dyes" in this chapter). When chlorophyll is withdrawn by the plant in the autumn, the leaves have yellow, brown and red tints because the carotene absorbs more weakly in the blue part of the spectrum while reflecting the "warmer" colored light (Figure 7.4).

When green plants are cooked as food, atoms of magnesium are replaced by those of hydrogen. This process causes the breakdown of the coloration revealing that delocalization of electrons is facilitated by the central, chelated, metal ion.

Chromophores

A chromophore is that part of a molecule responsible for its color. The color arises when a molecule absorbs certain wavelengths of visible light and transmits or reflects

FIGURE 7.4 (See color insert.) Color in autumn leaves. (Courtesy of http://www.hdwallpapersos.com)

others. The chromophore is a region in the molecule where the energy difference between two different molecular orbitals falls within the range of the visible spectrum. Visible light reaching the chromophore will be absorbed in a narrow band of frequencies by exciting an electron from a molecular orbital in the ground state to one in an excited state.

It is important to bear in mind that these electronic transitions are from one molecular orbital to another—these are not atomic spectra.

Chromophores are strongly associated with conjugated unsaturated systems of covalent bonds within molecules. In this chapter, on color in organic chemistry, many examples are presented of linear and cyclic molecules: β-carotene, a linear molecule, in "Saffron and Carotenoids: Yellow and Orange Dyes"; quinones, cyclic molecules in "Red Dyes from Henna, Dyer's Bugloss and Madder" and the purplish-blue dye, indigo in "Woad and Indigo."

The phenomenon of color in chromophores can be understood in terms of the electronic states of molecules. The atoms of the elements carbon, nitrogen and oxygen, with atomic numbers 6, 7 and 8, respectively, have partially filled p orbitals. The p orbital occupies a space around an atom, which has a "dumbbell" shape. When a double (or a triple) covalent bond is formed between two of these atoms, the "dumbbell" shapes overlap each other to form a molecular orbital, named a *pi* orbital. This orbital has a shape resembling a "sausage" above and below the plane of the bond. In a conjugated system of double and single bonds, the molecular *pi* orbitals of neighboring bonds merge into one, thus allowing the valence electrons in the p orbitals to range over the system, that is, the electrons are delocalized (see Figure 7.5).

Experimental measurements confirm that the carbon-to-carbon bond length in conjugated systems within molecules lies between the length of a single and a double bond. It should be noted that in other parts of this book many different organic substances are presented with delocalized systems of electrons in molecular orbitals, and each of these substances possesses a great deal of stability.

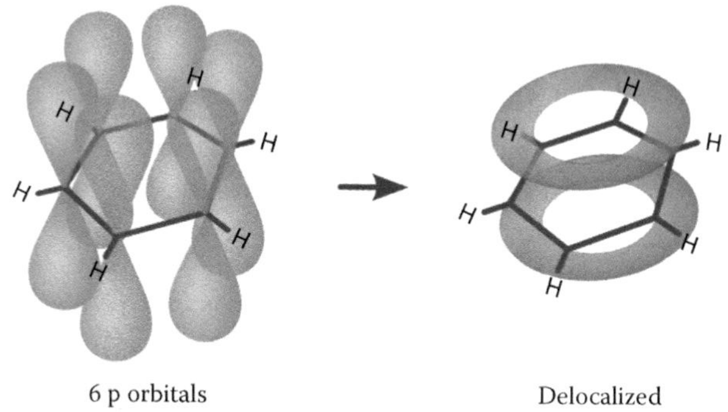

FIGURE 7.5 Orbital hybridization in benzene rings showing the *pi* bond. (Courtesy of Vladsinger.)

It has also been found empirically that the longer the carbon skeleton of the chromophore the lower the energy gap between the energy levels of *pi* orbitals. Thus, the longer chromophores, such as β-carotene, will absorb at smaller frequencies than shorter chromophores like benzene. Benzene, and other aromatic substances, such as naphthalene, appear colorless because they absorb in the ultraviolet region of the spectrum at higher frequencies, and hence, at higher energy than in the visible range. The chromophore of β-carotene absorbs in the blue end of the visible spectrum, transmitting light in the red and orange areas which are at lower frequencies, while the chromophore in benzene, and other aromatic substances, such as naphthalene and pyridine, absorb outside the visible spectrum in the ultraviolet, and thus, are colorless.

This effect can be seen also when a chromophore is changed and becomes smaller. Here, it will be instructive to compare the purplish-blue color of indigo with the white color of indigo white in the section on "Woad (*Isatis tinctoria*) and Indigo" presented later in this chapter. The color of the new chromophore will be due to light absorption in the ultraviolet at higher frequencies outside the visible region, shifting the apparent color perceived by reflected light from blue to white.

Of course, extending the length of a chromophore would have the opposite effect, so the inclusion of, for example, an azo functional group between two colorless benzene molecules will form an azo dye (and for more on azo compounds, see the sections on "Woad (*Isatis tinctoria*) and Indigo"and "Reversible Colors in Flowers, Berries and Fruit" in this chapter).

Molecular Interaction with Electromagnetic Radiation

Electromagnetic radiation interacts with the molecules of various substances in different ways, depending upon the energy of the radiation, the molecular structure of the compound and, to some extent, on the physical state of the compound, namely, whether it is a solid, gas, or liquid, or in solution. The general effects of the absorption of radiation are shown in Table 7.1 where the energy available from radiation increases from top to bottom of the table.

TABLE 7.1
Frequency Ranges of Various Light Sources

Frequency Range	General Effect
Microwave	Translation and rotation of a molecule
Infrared	Stretching vibrations in bonds
Visible	Color, due to visible light absorption by a chromophore
Near-ultraviolet	Absorption but no color—sometimes associated with subsequent fluorescence in the visible spectrum
Far ultraviolet	Structural change, dissociation of bonds

When ultraviolet or visible radiation strikes a molecule of an organic compound, the valence electrons in chemical bonds can absorb enough energy to jump to the next energy level. These are quantum effects, or step changes, therefore, the absorption of radiation occurs at a definite frequency relating to the energy gap of the transition involved. Of course, high-energy irradiation can cause so much energy to be absorbed that the bonds break to release free atoms or radicals. This occurs, for example, in the case of molecules of ozone in the upper atmosphere of the Earth (see also free radicals in the section on "Cocoa: Food of the Gods" in Chapter 4).

Ultraviolet Absorption Spectroscopy and Organic Chemistry

Molecular spectroscopy is a field of study which exploits each kind of electronic transition, whether the radiation involved lies in the infrared, visible, or ultraviolet regions of the spectrum. In the spectrum shown in Figure 7.6, one nanometer (nm) is one thousand billionth of a meter. The ultraviolet region lies in the range of wavelength from 400 to 280 nm.

Molecular spectroscopy is of interest to both organic chemists and biochemists who are concerned with the identification of large organic molecules, the identification of key chemical groupings in unknown or little understood organic molecules, and with quantitative analysis. Since speed of analysis and the convenient use of small samples are important considerations for these scientists, low-resolution ultraviolet absorption spectra in the liquid phase or in solution are especially valuable.

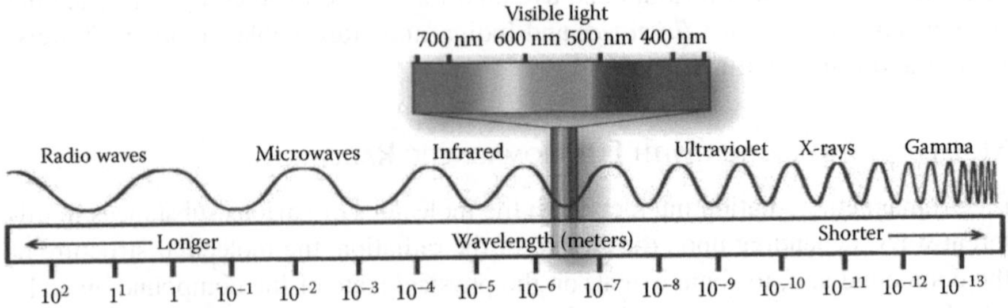

FIGURE 7.6 The electromagnetic spectrum of radiation. (Courtesy of Dr. Julie Lambert, Climate Science Investigations, South Florida.)

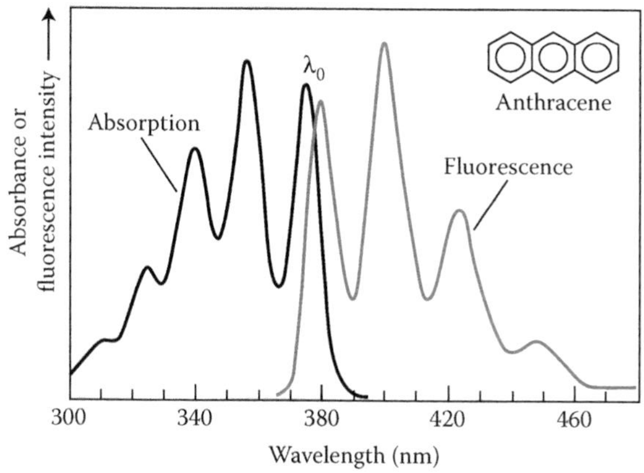

FIGURE 7.7 Absorption and fluorescence of anthracene. (http://www.reseapro.com Jablonski diagram archive.)

The effect of change of solvent has been widely studied but is not that significant, unless hydrogen bonding is involved. The nature of the solvent may slightly vary the absorption frequency of the solute and also the precise shape of the bands.

An illustration of the characteristic shape of an absorption spectrum of ultraviolet radiation is shown in Figure 7.7 for the compound, anthracene. Absorption is followed quickly by loss of the energy through fluorescence within the visible region in a mirror image, which relates to the similar physical structure of the molecule when it is in the higher energy state.

Low-resolution ultraviolet spectra have been of great benefit in organic chemistry, and complement information gained from studies of the structures of organic molecules arising from infrared absorption spectra and nuclear magnetic resonance absorption spectra.

Ultraviolet Absorption Spectroscopy and Molecular Structure

While saturated organic molecules seldom have transitions in the visible or near-ultraviolet regions of the spectrum, unsaturated molecules have low-lying excited states, which can be reached by absorption of radiation in this region.

The existence of a low-resolution visible or ultraviolet absorption spectrum can even be taken as evidence of unsaturation in molecules. For many unsaturated molecules, the absorption regions are defined and occur at recognizable, typical frequencies for a number of common structures in large organic molecules. There is considerable variation, both in the intensity and position of absorption, which is very helpful in identifying them in a given organic compound.

An increase in conjugation—that is alternate single and double bonds—is known to lower the frequency (and also raise the intensity) of the main absorption bands, as seen in Table 7.2, showing a range of unsaturated molecules.

Much is also known about the effect of functional groups on these absorption bands. Functional groups, which do not extend the degree of unsaturation of a

TABLE 7.2
Absorption Bands of Some Common Conjugated Compounds

Organic Substance	Wavelength in nm of Peak Absorption	Region
Ethane	180	Ultraviolet
1.3 Butadiene	217	Ultraviolet
Benzene	255	Ultraviolet
Phenol	280	Ultraviolet
Naphthalene	286	Ultraviolet
Paranitrophenol	320	Ultraviolet
Anthracene	375	Ultraviolet
β-carotene	450	Visible

molecule, have little effect though some, including hydroxyl and methyl groups, tend to lower the frequency of absorption in a reliable way. As an illustration, compare phenol with paranitrophenol in Table 7.2.

A very large bank of absorption data has been built up over time to help with the identification of an organic molecule or its structural features. Organic chemists regularly measure the ultraviolet absorption spectra of unknown or uncertain compounds (especially in the fields of polyene, aromatic, heterocyclic, steroid and carotenoid chemistry) in order to provide evidence for molecular structure. Even where the evidence is not conclusive, it may guide research in the chemistry of these compounds.

Green Plants and Limiting Climate Change

Some gaseous substances can absorb energy from the infrared part of the spectrum, which makes their covalent bonds stretch and vibrate even more. Only molecules comprised of different atoms, which are bonded together and differ in electronegativity, can interact with infrared radiation in this way, because of the dipolar nature of the chemical bonds.

Hence, oxygen and nitrogen, the principal natural constituents of the atmosphere by volume (21% and 78%, respectively), do not absorb infrared radiation, while carbon dioxide, water vapor, methane and nitric oxide do so quite strongly. Gases that absorb in the infrared are called "greenhouse gases" because they reduce the albedo effect of the Earth (energy radiated from the Earth back into Space).

It is, of course, the balance between incoming radiant energy received from the Sun and outgoing energy radiated from the Earth, which fundamentally drives the climate of the Earth such that any significant reduction in albedo results in global warming and climate change. A part of the section on "A Plant from the East Indies: Camphor" in Chapter 6, explores greenhouse gases (ghgs) in more detail.

Because of the sheer volume of carbon dioxide released into the atmosphere by man due to the burning of fossil fuels, greater initial attention has been given to measures to control carbon dioxide. However, the inadvertent release of more methane is a threat of some significance, because of the very high absorbance of infrared radiation by hydrocarbons. Some of this additional methane arises from natural sources through the effect of global warming itself on the permafrost in the tundra and on ocean sediments. Significantly more methane is released by energy-intensive farming practice, which provides protein from animal sources rather than from green plants. A switch in farming methodology toward green plants, and a reduction in demand for animal protein (arising from a change in diet in the wealthy nations of the world) would, ultimately, have beneficial effects on climate change.

Welcome and positive contributions to managing the effects of climate change would also arise from an increase in the capture or sequestration of carbon from the atmosphere through increasing photosynthesis by green plants brought about by

- Reversal of the scale of deforestation programs and
- Reduction in overgrazing of green plant habitat by farmed animals.

Other approaches to carbon capture involve direct physical intervention by man to remove waste carbon dioxide from the exploitation of fossil fuels in the energy-generation industry by liquefying the gas under pressure and forcing it into underground reservoirs in deep mines or redundant oil and gas fields.

Reduction in dependence on fossil fuels by the use of renewable sources of energy would be advantageous in respect of climate change, though there are distinct practical and political issues associated with this idea. However, the use of renewable raw materials and the recycling of biodegradable substances in the long term would be of considerable value.

A NOTE ON CARBON ACCOUNTING AND CARBON NEUTRALITY

With comprehensive, binding, international agreements on climate change still somewhat elusive, sporadic regional and local initiatives have arisen in enlightened government and business circles.

Though ambitious, some institutions have made public statements about their aim to achieve a zero-carbon footprint or carbon neutrality in the pursuit of their activities. At the very least, such aims offer a strong statement of environmental commitment appealing to many individuals whether electors, or customers, or those involved in the supply chain.

Carbon neutrality is a condition in which net ghg emissions associated with an organization or a product are zero. Ideally, in order to make this effective and meaningful, an agreed system of carbon credits would have to be introduced worldwide, which would be typically used to counterbalance or compensate for carbon dioxide emissions elsewhere in a process. Such offsetting would have to be based upon an internationally acceptable unit of measurement, or CO_2 equivalent. Very probably, such carbon credits could be legitimately traded between organizations in a regulated market.

However, meaningful, widely accepted standards with which to compare the effects of one human activity with another do not yet exist. There is still no universally accepted unit of carbon accounting although an expression of this is in use. A CO_2 equivalent for a ghg may be regarded as the amount of CO_2 in metric tons which would have the same global warming potential when measured over a 100-year timescale.

Although governments in Japan or Sweden, for instance, have set energy-efficiency targets, one clear example of resolve in carbon accounting is provided by the British Government which, in 2008, passed the UK Climate Change Act. This act of parliament set up the Department of Energy and Climate Change, and established a legal obligation to report, annually, progress in the United Kingdom toward the meeting of carbon targets. In fact, the overall carbon target is to reduce ghg emissions by at least 80% by 2050. Intermediate carbon targets have been set covering the period 2008–2027. The United Kingdom has made a commitment to halve ghg emissions by 2027 relative to 1990. Where emissions rise in one sector of the economy, the UK Government will look to reduce them in other sectors by a corresponding amount. In 2014, emissions were reported to be down by a quarter since 1990. On the basis of current policy, the UK Government appears to be on track to achieve a reduction of one-third by 2020. However, over the next 10 years, technologies will have to be developed and deployed to achieve the overall target of a reduction of 80% by 2050. The United Kingdom is attempting to move to a lower carbon, more sustainable economy, which is far less dependent on irreplaceable fossil fuels. Much improved fuel efficiency in the space heating of buildings is expected to make a large contribution.

QUESTIONS

1. What are lone pairs of electrons, and in what atoms are they present in organic molecules? Explain the role of lone pair electrons in chemical bonding. Give examples of linear and cyclic organic molecules which can act as ligands.
2. Provide three examples of symbiosis in the plant kingdom, and explain the basis of the relationship between each pair of organisms.
3. Explain why green plants are of critical significance in arresting climate change.
4. Why do green plants appear green?
5. Explain the difference between primary and secondary metabolites in plants, and provide examples of each kind of compound.
6. By reference to the theory of molecular orbitals, explain how conjugated systems in organic molecules can give rise to chromophores.
7. β-carotene, presented in the section on "Saffron and Carotenoids: Yellow and Orange Dyes," quinones, found in the section on "Red Dyes from Henna, Dyer's Bugloss and Madder" and the alkaloids, indigo and indigo

white, are described in the section on "Woad (*Isatis tinctoria*) and Indigo." Explain how the length of the chromophore influences the frequency of light absorbed by the molecular orbital, and hence, the color of the compound.

8. Account for dipolar bonds in covalent molecules. Many of the gases in the atmosphere of the Earth are transparent to infrared radiation but two, in particular, absorb strongly, reducing the albedo effect and are known as "greenhouse gases." Which are they?

   Explain why many pollutant gases (methane, oxides of nitrogen and carbon hydrogen fluorides) are very influential as "greenhouse gases," and indicate the main sources of emissions.
9. Give an account of the environmental disadvantages of excessive exploitation of green plants in some parts of the world, examples being deforestation and overgrazing of grasslands. Allow for the fact that, in certain countries, man-made monocultures of crops are grown extensively.
10. What is zero-carbon accounting? Give a balanced account of the pros and cons of this worthy and ambitious approach to limiting climate change.
11. What are the environmental effects of the rising levels of emissions of carbon dioxide which are being released into the atmosphere of the Earth by the activities of man? Describe natural and artificial means of carbon capture which may address the issue.

REFERENCES

UK Government. 2014. *Reducing the UK's Greenhouse Gas Emissions by 80% by 2050*. UK Department of Energy and Climate Change, UK.
R. H. Whittaker. 1975. *Communities and Ecosystems*. Macmillan.
R. H. Whittaker, Ed. 1978. *Classification of Plant Communities, Handbook of Vegetation Science*. Kluwer Academic Publishers.

SUGGESTED FURTHER READING

C. J. Balhausen. 1962. *Introduction to Ligand Field Theory*. McGraw-Hill.
J. N. Murrell, S. F. A. Kettle, J. M. Tedder. 1965. *Valence Theory*. John Wiley.
A. Streitwieser Jr. 1961. *Molecular Orbital Theory for Organic Chemists*. John Wiley.
Zero Carbon Britain: Rethinking the Future. 2014. Centre for Alternative Technology (CAT). Available at http://www.zerocarbonbritain.org.

SAFFRON AND CAROTENOIDS: YELLOW AND ORANGE DYES

Abstract: The golden-colored essence of saffron, namely crocin, has always been a highly prized, luxury item fostering trade over millennia, and remains one of the most expensive plant commodities by weight.

Natural products chemistry

- Crocin—a terpenoid dye
- Used in food flavoring and as a food colorant
- β-carotenoid and other terpenoid colorants derived from plants
- The allyl functional group.

Curriculum chemistry

- Carotenoids which contribute to photosynthesis in a different part of the visible spectrum to chlorophyll and to autumn color in plants, and are present in tomatoes and root crops such as carrots
- Human health and vitamin A
- Retinol and retinal.

SAFFRON: THE PLANT

The saffron crocus, *Crocus satirus* (Figure 7.8), is unknown in the wild, having been developed by selective plant breeding and propagation through vegetative

FIGURE 7.8 **(See color insert.)** Attractive flowers of the Saffron crocus. (Courtesy of Dobies of Devon, UK; flower and seed merchants.)

multiplication of the *Crocus cartwrightianus*, which is native to the Greek island of Crete. Documented history of cultivation spans millennia.

Dried stigmas of the plant are used in food flavoring as seasoning and as a food colorant.

The stigma is the female part of the reproductive system in a flower. It receives pollen and it is on the stigma that the pollen grain germinates. The red structure in the picture is the stigma of the crocus flower.

A NOTE ON SAFFRON EXTRACT: AN EXCLUSIVE, LUXURY ITEM TRADED FOR MILLENNIA

Given the labor intensive methods involved in cultivation, separation of the stigmas and extraction of the golden-colored essence, it comes as no surprise to learn that saffron has always been a highly prized, expensive, luxury item, which has fostered trade over millennia. In fact, it may take up to 250,000 stigmas to produce one pound of saffron, therefore, saffron remains one of the most expensive plant commodities by weight.

Owing to high costs, saffron has been cultivated in the West nearer to its point of use. The cultivation of saffron was introduced into England in 1530 mainly in the drier eastern part of the country. The town of Saffron Walden in the county of Essex owes its name to the plant. Saffron continues to be grown commercially in the eastern county of Norfolk. By 1730, the Pennsylvania Dutch, settlers from central Europe, had established cultivation of saffron in eastern Pennsylvania in the United States, and production continues today around the city of Lancaster. However, today, much of the world crop is produced in Iran.

Crocin: The Extract from Saffron

Crocin (Figure 7.9) is found in the flowers of the crocus family, and is the substance responsible for the golden color of saffron. Crocin is a carotenoid—a colored pigment ultimately derived from terpene. Crocin has a deep red hue but, when dissolved in water, it forms a solution with the characteristic golden orange color. It has a very large molecular mass ($C_{44}H_{64}O_{24}$), and is a crystalline solid at room temperature. The central part of the molecule consists of four isoprene units.

Crocin has been shown to be a potent antioxidant and has also been shown to prevent the proliferation of cancer cells.

Carotenoids and Autumnal Colors

Carotenoids are pigments ultimately derived from terpene. The breakdown of chlorophyll and other compounds in the leaves of deciduous plants and trees during the

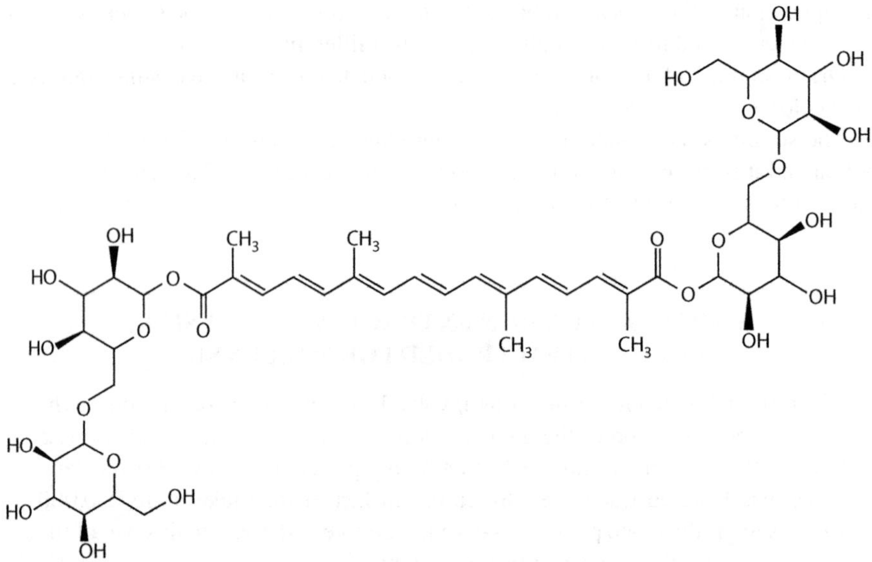

FIGURE 7.9 Structure of crocin.

autumn produces sugars which are stored over winter in the root system. The process of breakdown creates a natural palette of reds, oranges, yellows and browns, as carotenoid substances are left behind and exposed.

Without carotenoids in the leaves of plants, episodes of excess light energy could destroy proteins and membranes. In addition, some scientists believe that carotenoids can act also as regulators of developmental responses in plants.

Naturally occurring carotenoids obtained from plants, which are used for dyeing include

- Lutein from nettles, French marigolds and many other plants
- Bixin from annatto obtained from the seeds of the achiote tree (*Bixa orellana*)
- Crocin from the stigma of the flowers of the saffron plant, *Crocus satirus.*

Carotenoids belong to a family of brightly colored natural compounds found in flowers and fruits. They are classified into two families: carotenes and xanthophylls. The former contain carbon and hydrogen atoms, whereas the latter also contain oxygen atom(s) in the structures. All carotenoids have a long hydrogen carbon chain with alternating single and double bonds. The molecular skeleton of carotenoids contains 40 carbon atoms arising from eight multiples of the naturally occurring building block, isoprene (see Figure 7.10). They are, therefore, tetraterpenoids.

Their color in the visible spectrum of light is due to the large number of double bonds and associated electron delocalization.

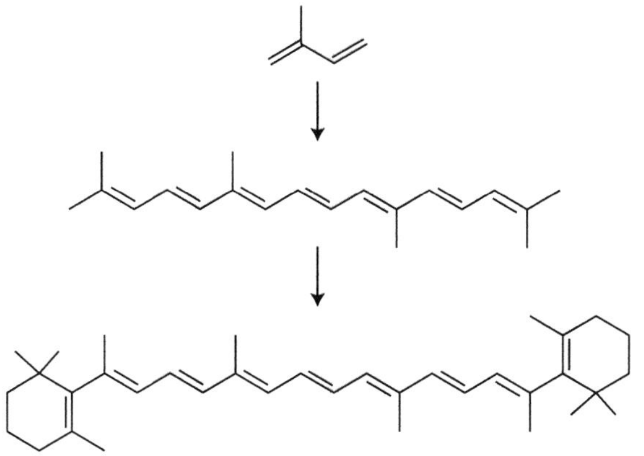

FIGURE 7.10 Beginning from an isoprene building block, the pathway to β-carotene.

CAROTENOIDS: HISTORICAL NOTE

The earliest studies on carotenoids date from the middle of the nineteenth century. Carotene was isolated in 1831. In 1837, Berzelius gave the name xanthophylls to the yellow pigments he obtained from autumn leaves. By the 1860s, lutein had been identified. The first separation of the two main types of carotenoids, carotenes and xanthophylls, was achieved through chromatography in 1910 by Willstater and Mieg who established the empirical formula of $C_{40}H_{56}$ for β-carotene and $C_{40}H_{56}O_2$ for xanthophylls. It was not until the 1930s that Karrer elucidated the chemical structures of carotenoids, and also demonstrated that carotenoids are transformed into vitamin A in the human body.

Carotenoids, Carotenes and Xanthophylls

Two main types of carotenoids are found in nature: carotenes and xanthophylls.

Carotenes

Carotene-rich foods are often orange in color, for example, carrots, sweet potatoes, winter squash, spinach, kale and fruits like cantaloupes and apricots.

β-Carotene

β-Carotene (Figure 7.11) is a very important member of the carotenoid family, and is the precursor to vitamin A.

Lycopene

Lycopene (Figure 7.12) with molecular formula, $C_{40}H_{56}$ is a bright red carotene pigment from tomato species and other red fruits and vegetables such as red

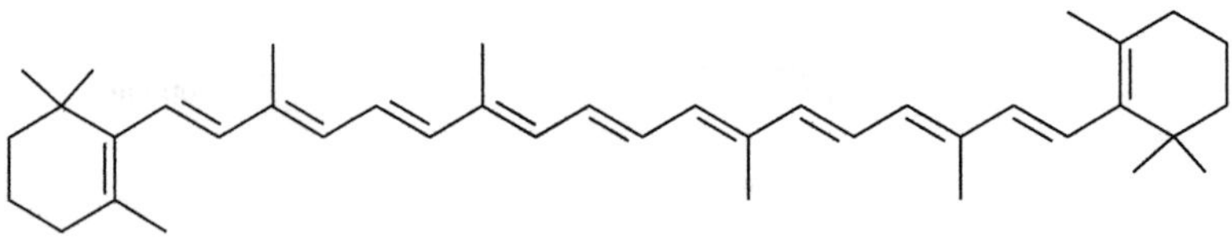

FIGURE 7.11 The chemical structure of β-carotene.

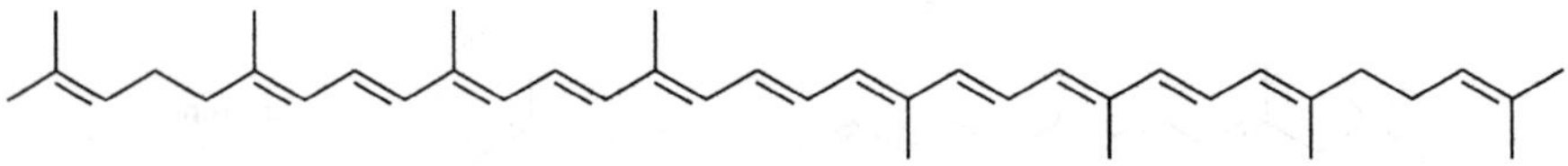

FIGURE 7.12 The molecular structure of lycopene.

carrots, watermelons and papayas. Due to the presence of 11, conjugated, double bonds, lycopene is deep red and, as an unsaturated compound, has the properties of an antioxidizing agent. Normally, lycopene is used as food coloring in the food industry.

Isolation and Extraction

Use of supercritical CO_2 is a well-established method for extraction of β-carotene and lycopene from waste tomato paste and even watermelon (for more about supercritical CO_2 see the section on "Coffee: Wake Up and Smell the Aroma" in Chapter 4). CO_2 is often chosen for its supercritical fluid extraction method, as it has a low critical temperature, is nonflammable, and can be obtained at low cost and in high purity. For example, the skins and seeds of dried, ground-up, ripe tomatoes are extracted as follows: the extraction conditions are 2,500–4,000 psi pressure and 40–80°C for 30 minutes. Lycopene is more readily dissolved in supercritical CO_2 than other carotenes. At 4,000 psi and 80°C, the extraction product contains up to 65% of lycopene and 35% of β-carotene.

Xanthophylls

Crocin

Crocin is mentioned briefly, as it is described earlier in this chapter, and is shown in Figure 7.9. Crocin is the xanthophyll carotenoid found in the flowers of the crocus family, and is the substance responsible for the golden color of saffron.

Lutein

Lutein belongs to the xanthophyll family of carotenoids. Again, the molecular structure is made up from isoprene building blocks in the form of a tetraterpenoid compound, but with oxygen atoms also present (see Figure 7.13). Lutein is obtained from green leafy vegetables, such as spinach and kale. Lutein is also responsible for the yellow color of egg yolk, chicken skin and fat. The very name, lutein, comes from Latin meaning "yellow."

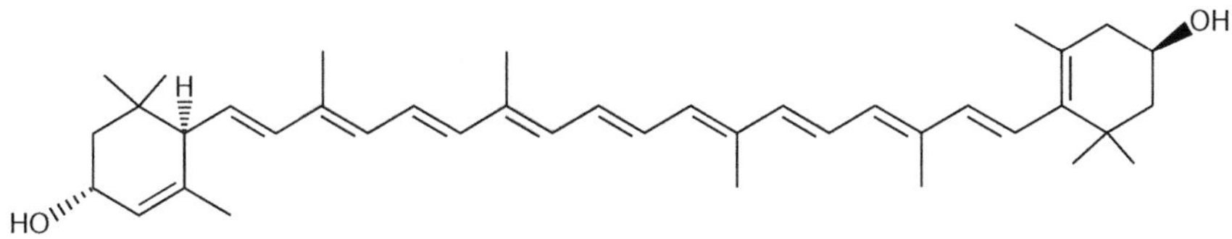

FIGURE 7.13 Lutein.

According to the American Optometric Association, studies show that lutein reduces the risk of chronic eye diseases, including age-related macular degeneration (AMD) and cataracts.

AMD is a painless eye condition that generally leads to the gradual loss of central vision. The macula is that part of the retina at the back of the eye, which is responsible for central vision. As the name of the condition suggests, loss of vision is gradual over many years, and usually occurs with aging. AMD is a leading cause of visual impairment and is most common in people over 50 years of age. It is estimated that one in every 10 people over 65 years of age has some degree of AMD.

Because of beneficial effects on eye health, small amounts of lutein added to daily diet are recommended, as human beings cannot produce this compound in the body.

Lutein has one β-ring and one ε-ring. Again, it is the presence of the long, conjugated, double bonds (the polyene chain), which cause its distinctive light-absorbing properties. The polyene chain is susceptible to oxidative degradation by light or heat, and is chemically unstable in acids. It is noteworthy that the position of the double bond in lutein involves a chemically reactive, allylic hydroxyl, functional group.

The Allyl Functional Group

The allyl functional group (Figure 7.14) has the formula $H_2C{=}CH{-}CH_2R$ where R represents the remainder of a molecule. The allyl functional group consists of a methylene bridge ($-CH_2-$) attached to a vinyl group ($-CH{=}CH_2$).

The site on the saturated carbon atom is known as the "allylic position." A group attached at this site infers the activity of this allylic compound. Thus, allyl alcohol, $CH_2{=}CHCH_2OH$, has an allylic hydroxyl group. Since the C–H bonds in an allylic functional group are weaker than the C–H bonds located in ordinary tetrahedral carbon centers, such as in methane, for example, the allylic C–H bonds exhibit more reactivity. This heightened reactivity has many practical applications in organic chemistry.

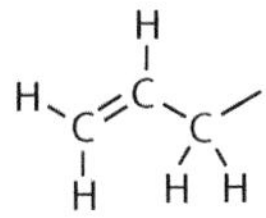

FIGURE 7.14 The allyl group.

Many allylic compounds have a lachrymatory effect on human beings (for more on allyl compounds, see also the section on "Garlic and Pungent Smells" in Chapter 3.)

Human Health: Vitamin A, Retinol and Retinal

Vitamin A is the collective name for a number of unsaturated nutritional organic compounds, which include retinol, retinal and several other carotenoids, among which β-carotene is the most significant. Vitamin A has multiple functions in that it is important for

- Growth and development
- Maintenance of the immune system
- Good vision.

β-Carotene is cleaved by the action of an enzyme in the human body to form retinol (Figure 7.15).

Retinol functions as a form of storage for the vitamin in the body. Chemical reduction of the alcohol functional group readily occurs in the body to give the visually active and corresponding aldehyde, retinal. Vitamin A is needed by the retina of the eye in the form of retinal, which influences low-light and color vision.

You will appreciate from Figure 7.11 that retinol is a diterpenoid compound. It has two isoprene units in the middle of the molecule. Many different geometric isomers of retinol and retinal arise from the *trans* or *cis* configurations in four of the five double bonds found in the polyene chain. Some *cis* isomers carry out essential functions in visual acuity. The 11-*cis*-retinal isomer, for example, is the chromophore of rhodopsin—the photoreceptor molecule in vertebrate animals. Vision depends upon light-induced isomerization of the chromophore from the *cis* to the *trans* isomer. This change of conformation activates the photoreceptor molecule. Night blindness, an inability to see well in dim light, is one of the earliest signs of vitamin A deficiency followed by decreased visual acuity. In 1967, George Wald won the Nobel

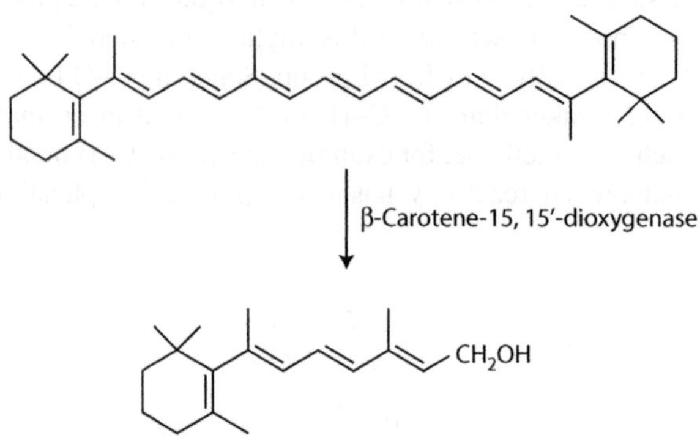

FIGURE 7.15 Conversion of β-carotene to retinol.

Prize for his work with retinal pigments, which led to the understanding of the role of vitamin A in vision.

QUESTIONS

1. The allyl functional group is very reactive. Illustrate the breadth of usefulness to organic chemistry of compounds containing the allyl group.
2. Explain why ultraviolet spectroscopy is a much more important tool for studying carotenoids than infrared spectroscopy.
3. Explain the difference between the carotene carotenoids and the xanthophyll carotenoids, and illustrate your answer with exemplar compounds.
4. Choose one tetraterpenoid, and illustrate the nature and prevalence of geometric isomerism.
5. Explain why the addition of one or more functional groups containing a nitrogen atom or an oxygen atom to a molecule of a dye or pigment can change the color of the chromophore.

REFERENCES

K. Akhtari, K. Hassanzadeh, B. Fakhraei, N. Fakhraei, H. Hassanzadeh, S. A. Zarei. 2013. A density functional theory study of the reactivity descriptors and antioxidant behavior of Crocin. *Computational and Theoretical Chemistry* 1013:123–129.

D. G. Chryssanthi, F. N. Lamari, G. Iatrou, A. Pylara, N. K. Karamanos, P. Cordopatis. 2007. Inhibition of breast cancer cell proliferation by style constituents of different Crocus species. *Anticancer Research* 27(1A):357–362.

J. Escribano, G. L. Alonso, M. Coca-Prados, J. A. Fernandez. 1996. Crocin, safranal and picrocrocin from saffron (*Crocus sativus L.*) inhibit the growth of human cancer cells in vitro. *Cancer Letters* 100(1–2):22–30.

A. Rubio-Morago, R. Catillo-Lopez, L. Gomez-Gomez. 2009. Saffron is a monomorphic species revealed by RAPD, ISSR and microsatellite analyses. *BMC Research Notes* 2:189.

SUGGESTED FURTHER READING

R. Cooper, G. Nicola. 2014. *Natural Products Chemistry: Sources, Separations and Structures*. CRC Press, Taylor & Francis Group.

D. B. Gregg. 1974. *Agricultural Systems of the World*, 1st edition. Cambridge University Press.

T. Hill. 2004. *The Contemporary Encyclopaedia of Herbs and Spices; Seasonings for the Global Kitchen*. Wiley.

WOAD (*ISATIS TINCTORIA*) AND INDIGO

Abstract: Indigo, a purplish-blue dye obtained from the plant, Woad, is among the oldest of natural dyes, and is used in textile dyeing, body art and printing.

Natural products chemistry

- Indigo—a purplish-blue dye
- Indigo—a naturally occurring alkaloid
- Indigo produced as a synthetic dye in the modern world
- Indigo as a pigment in body painting and in the dyeing of textiles.

Curriculum chemistry

- The emergence of the modern synthetic dye industry
- The production and use of synthetic azo dyes.

Indigo through the Ages

Records indicate that even as late as the 1980s in countries of the southern Arabian Peninsula, people applied indigo as a healing treatment to their skin. They found that wrapping an indigo cloth, coated in beeswax and oil, around a wound was an effective antiseptic procedure.

Indigo became so valuable as a textile dye that it was traded extensively from East to West, and ultimately drove the early development of the industrialization of synthetic organic chemistry during the middle to late 1860s in Europe through the leadership of such scientific giants as William H. Perkin and Adolf von Baeyer. While there are plenty of historical accounts of the use of indigo, many of which list countless ailments that indigo has been reputed to cure, it is the pigment and coloring properties of the alkaloid which are examined here.

HISTORICAL NOTE ON INDIGO

Indigo is among the oldest of dyes used for textile dyeing and printing. In many Asian countries (such as India, China, Japan and south eastern Asia), indigo has been used as a dye over thousands of years. The dye was also known to ancient civilizations in Mesopotamia, Egypt, Britain, ancient Persia and Africa. As a highly valuable commodity, dyes from India spread west along established ancient trade routes. It was via Arab merchants that indigo came to the attention of the Roman Empire where it was considered a luxury item. Although in the first century AD the Roman historian, Pliny the Elder, refers in texts to the common use of indigo and woad by the Gallic tribes of northern Europe, indigo had been established in use since the Bronze Age (2,600–800 BC) particularly in Celtic Britain. Indigo was known as woad since it was obtained from the native wild plant known as Woad, *Isatis tinctoria*.

When the sea route from Europe to India was opened in the late fifteenth century, direct trade flourished since dangerous land routes and tax duties imposed by the Persians and other intermediaries were avoided. Ease of import of indigo through ports in Portugal, The Netherlands and England meant that use of the dye grew significantly in Europe.

Rising European demand for indigo stimulated the creation of indigo plantations in other parts of the world. Indigo became a major crop in both Jamaica and in South Carolina where enslaved Africans and African Americans met the need for labor. Indigo developed into a major export crop in the plantations of colonial South Carolina, which tended to reinforce dependence on slavery during the eighteenth century.

Owing to its high value as a trading commodity, indigo was often referred to as blue gold.

INDIGO FROM PLANTS

Most of the indigo obtained from natural sources came from plants in the genus *Indigofera,* which are native to the tropics. The primary commercial species in Asia is *Indigofera tinctoria.*

In Celtic Britain, where the dye was used as a form of war paint by native tribes, indigo was known as Woad since it was obtained from the native wild plant, Woad, or *Isatis tinctoria.*

The molecular structure of indigotin, commonly known as indigo, is shown in Figure 7.16.

Indigotin is directly related to the important, natural, building block, indole—the molecular structure of which is shown earlier in the book in Figure 4.11. Indole is also described in various sections—see, in particular, "Africa's Gift to the World" in Chapter 2 and "Maca from the High Andes of South America" in Chapter 4.

Indole is relatively stable, chemically, due to a high degree of electron delocalization above and below the plane of the molecule. In turn, the chemical stability of the indole building block markedly influences the chemistry of indigo, which is remarkably unreactive.

An interesting feature of indigo requires some explanation. The molecule is planar, so the possibility of geometric isomerism has to be considered, yet only the *trans* isomer has ever been isolated. This phenomenon is understood in terms of steric hindrance, especially between the neighboring oxygen atoms, which destabilizes the

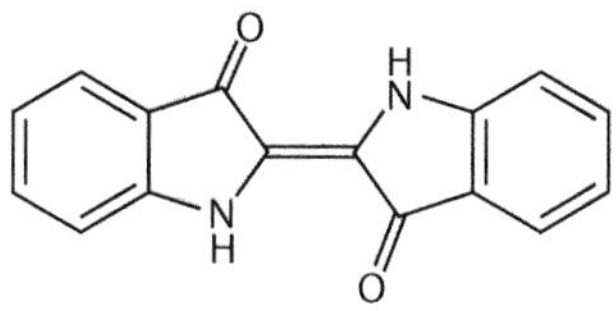

FIGURE 7.16 The structure of indigo, an alkaloid dye.

cis isomer. More on geometric isomerism is presented in the sections on "Saving the Pacific Yew Tree" in Chapter 2 and "Cannabis and Marijuana" in Chapter 5.

Extraction of Indigo

The precursor molecule to the chemical, indigo, is called indican. Indican is a colorless organic compound, soluble in water, which occurs naturally in *Indigofera* plants. Indican is obtained by processing leaves of the plant, which contain a small percentage by weight of the compound. Indican is a glycoside (see also the section dealing with polysaccharides, "Asian Staple: Rice" in Chapter 3). In fact, the chromophore in many dyes and pigments is often chemically bound to a glucose moiety through a glycoside bond. As a glycoside, indican will undergo hydrolysis to yield glucose and a molecule known as indoxyl in which the chromophore is located. Upon exposure to air, indoxyl is readily oxidized to indigo.

Thus, the extraction process firstly involves soaking the leaves in water in the presence of air to release the blue dye as a precipitate. This material is then ground up with alkali and pressed into cakes from whence it is dried and powdered. This powder can be mixed with various other substances to produce different shades of blue and purple (Figure 7.17).

Physical and Chemical Properties of Indigo

Indigo is a dark blue crystalline powder, which sublimes at 390–392°C. It is insoluble in many organic liquids such as water, alcohol and ether. The very low solubility of indigo in organic solvents is attributable to both intramolecular and intermolecular hydrogen bonding.

The molecule absorbs light in the orange part of the visible spectrum (λ_{max} = 613 nm) and, therefore, the compound appears to the eye in the complementary color, purplish blue. The chromophore (see Glossary) in indigo has been shown to be the central part of the molecule without influence from either of the two phenyl

FIGURE 7.17 **(See color insert.)** Indigo from natural sources. (The historical collection of dyes at the Technical University of Dresden, Germany.)

groups. Similar to other conjugated compounds, the deep color arises from the delocalization of electrons above and below the series of three adjacent double bonds within the planar molecule. In this regard, it is not surprising that indigo and some of its derivatives possess the properties of organic semiconductors when deposited in thin films.

Indigo and Dyeing Textiles

When indigo first became widely available in Europe in the sixteenth century, it was found to be difficult to manage as a textile dye because it is not soluble in water. In order to overcome this difficulty, indigo was subjected to chemical reduction involving the nitrogen-to-oxygen double bonds. The reduction product is commonly known as indigo white or leuco indigo (Figure 7.18).

The process is as follows: when a submerged fabric made from cellulose is removed from the dye bath, indigo white quickly combines with oxygen from the air and reverts to the insoluble, intensely colored indigo, which is lodged in the weft and warp of the cellulose material by hydrogen bonding, since the cellulose polymer has many hydroxyl bonds (see cellulose in the section on "Global Aloe" in Chapter 6, and polysaccharides in the section on "Asian Staple: Rice" in Chapter 3). Relatively weak adherence through hydrogen bonding explains why the color is not fast, that is, it fades slowly, especially under the influence of regular washing of the fabric.

Levi Strauss and Wrangler Jeans

Indigo is primarily used as a dye for the cotton yarn used in the production of denim cloth in blue jeans. On average, a pair of blue jean trousers requires 3–12 g of indigo. According to Steingruber writing in *Ullmann's Encyclopedia of Industrial Chemistry*, about 20 million kilogram of indigo are produced annually and applied mainly in the dyeing of blue jeans. Small amounts are used for dyeing wool and silk. Synthetically produced indigo is used today for the dyeing of modern blue jeans made by such well-known companies as Levi Strauss and Wrangler.

Indigo is also permitted for use as a food colorant by the Food and Drugs Administration in the United States.

"Lincoln Green" Worn by Robin Hood

In medieval England, "Lincoln Green" was a favorite dyestuff for coloring woolen fabric. "Lincoln Green" was celebrated in the legendary story of the nobleman-turned fugitive, Robin Hood, who operated from Sherwood Forest in Nottinghamshire to do good for many disadvantaged serfs and villains of the time. Garments colored in

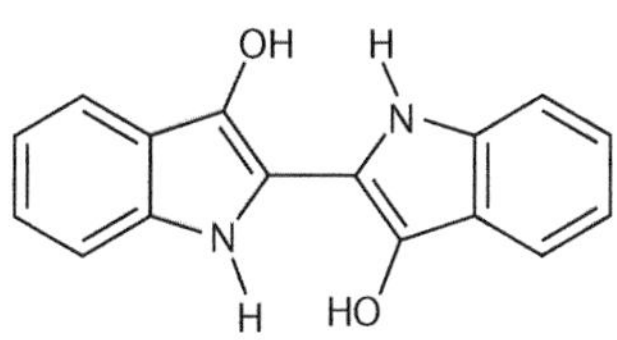

FIGURE 7.18 Indigo white or leuco indigo.

"Lincoln Green" would certainly have provided good camouflage within the forest for the band of followers led by Robin Hood.

The color, "Lincoln Green," was formed by blending purple-blue indigo with a yellow dye from the flowering wayside and garden plant, Tansy, *Tanacetum vulgare*.

HISTORICAL NOTE ON TANSY

Tansy has a long history of use. The ancient Greeks cultivated tansy for its use as an insect repellent. In modern times, tansy has been planted in potato fields to discourage the predatory Colorado beetle. Dried tansy has also been burned as incense on account of the camphor it contains. However, soaking the leaves and flowers in water and boiling the solution produces, on cooling, a mustard-yellow extract effective in dyeing wool.

The cyclic monoterpenoids, thujone, myrtenol and camphor, may be extracted as a volatile oil from the yellow flowers of Tansy, a perennial, herbaceous plant of the aster family. These cyclic monoterpenoids all have the same molecular formula, $C_{10}H_{16}O$, and thus, are considered to be examples of isomers (see the section on "Saving the Pacific Yew Tree" in Chapter 2). For more on cyclic terpenoids, see the sections on "Frankincense and Myrrh" and "A Plant from the East Indies: Camphor" in Chapter 6 and "Saffron and Carotenoids" in this chapter.

A HISTORICAL NOTE: THE MODERN SYNTHETIC DYE INDUSTRY REPLACES NATURAL DYES AT THE START OF THE TWENTIETH CENTURY

Aromatic amines can be readily oxidized to a variety of compounds which possess intense color. The most famous of these synthetic dyes was discovered by accident in 1856 by the English chemist, William H. Perkin, who was trying to find a means of synthesizing quinine while studying at the Royal College of Chemistry in London. Perkin's first experiments yielded only a black solid mess, which is often the result of a failed organic synthesis. But how poor had his experimental technique really been? Perkin noticed that the black products were consistently reproducible. When cleaning his reaction flask with alcohol, the keen-eyed Perkin saw the significance of the purple-mauve solution he had produced. Perkin obtained a patent from the British Patent Office for this colored material, which he named mauveine, having also found that the substance was effective in dyeing silk and other textiles. Shortly afterward, he set up a factory to produce the dye which became a commercial success.

Beyond the profound significance of this discovery as a technological breakthrough were the simple facts that (i) he used aniline as the starting material, which is readily available from coal, and (ii) the chemistry of the process is straightforward.

Perkin had found that when aniline is exposed to a strong oxidizing agent, for example, acidified potassium dichromate solution, mauveine was produced along with a number of other oxidation products such as p-benzoquinone (or 1–4 benzoquinone), a bright yellow solid.

However, the molecular structure of mauveine proved difficult to determine and, in fact, was only finally identified in 1994 by Meth-Cohn and Smith. One type of mauveine molecule is known as mauveine A. It has the molecular formula, $C_{26}H_{23}N_4^+X^-$, and incorporates two building blocks of aniline within a fused five-ring structure.

There was a drive to reduce dependence upon natural sources of indigo. Advances in synthetic organic color chemistry continued from the time of William Perkin through the studies of synthetic dyes by the famous German chemist, Adolph von Baeyer. He researched the synthesis of indigo and determined its chemical structure in 1870 and first described a synthetic method for producing indigo in 1878. However, the process was difficult in practice and still depended on plant extracts as source material. Therefore, a search for alternative chemical routes continued at the chemical laboratories of the German companies, Badische Anilin und Soda Fabrik (BASF) and Hoechst. Eventually, an alternative process was identified, which had the potential to permit synthetic production of indigo on an industrial scale, starting from readily available aniline. By 1897, BASF had developed a commercial manufacturing plant. Von Baeyer was awarded the Nobel Prize for chemistry in 1905 partly in recognition for his work on indigo. Today, modern production of indigo is based on naphthalene as the feedstock.

Within just a few years of their discovery, synthetic dyes had almost completely replaced natural dyes because they

- Offered a kaleidoscopic range of color and shade
- Possessed a brilliant, intense presence
- Made available colors which would act on a wide range of fabrics
- Did not wash out easily
- Did not fade on exposure to natural light
- Presented opportunities to develop reliable color reproduction.

Nevertheless, human curiosity ensures, even well into the twenty-first century, that interest in the potential of natural dyes remains undimmed.

Azo Dyes

In the modern world, the most common synthetic coloring agents used in textiles, in pharmaceutical products, in foods and in cosmetics are known as azo compounds.

Kekule, working in Germany in 1866, was the first chemist to identify the azo functional group, which comprises the double-bonded nitrogen atoms, as shown in Figure 7.19.

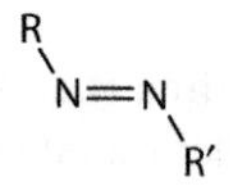

FIGURE 7.19 The generalized formula of an azo compound.

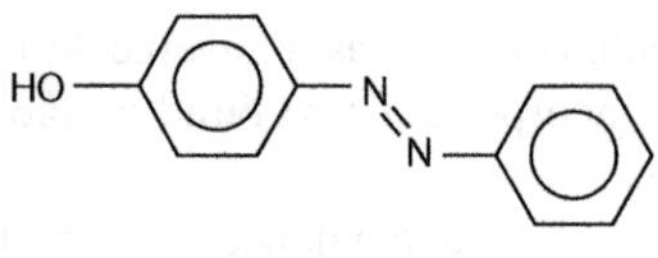

FIGURE 7.20 The molecular structure of an aromatic yellow azo dye.

The azo group may be bonded to aromatic groups, which, therefore, indicates that these molecules have extended conjugated systems that act as chromophores and impart vibrant and dense color, usually red, orange or yellow. An example is shown in Figure 7.20. Electron delocalization also imparts stability to aromatic azo compounds.

One of the earliest commercial textile dyes produced was called chrysoidine, a yellow compound, which has been in use for dyeing wool since 1875. It was synthesized from aniline and m-phenylenediamine. The synthesis of azo dyes named Congo Red and Bismarck Black soon followed in the late 1800s.

Azo dyes can be made in the laboratory in a three-step process. Firstly, phenylamine is added to nitrous acid in the presence of hydrochloric acid, which produces benzenediazonium chloride. Sodium phenoxide is then produced separately by dissolving phenol in sodium hydroxide solution. When the two solutions are added together at a low temperature just above the freezing point of water, the benzenediazonium ion, an electrophile, combines with the phenate ion, a nucleophile, to yield a yellow precipitate—the azo compound. This reaction is shown in Figure 7.21.

In addition to dyeing textiles, azo dyes are utilized also as pigments in paints and as acid/alkali indicators in chemical analysis in the laboratory. Methyl red, for example, described in the section on "Reversible Colors in Flowers, Berries and Fruit" in this chapter, is an azo dye.

A widespread application of azo dyes is as a colorant in the food processing industry. In the European Union, many azo food colorants are listed for brevity under what is known as an E number. For example, the azo dye, tartrazine, a

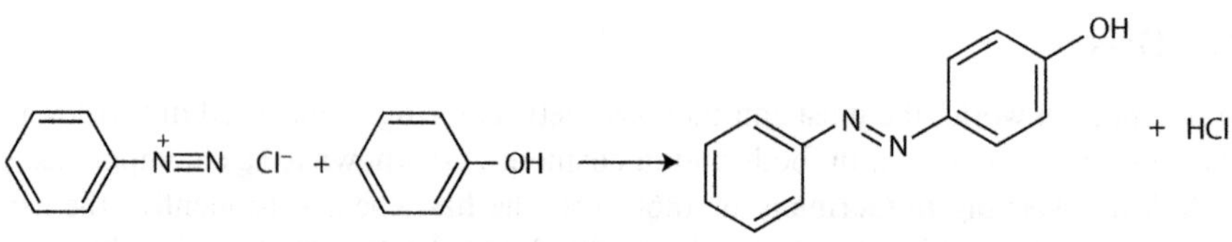

FIGURE 7.21 Coupling of benzenediazonium chloride and phenol.

well-known food colorant, is identified as E102. Tartrazine imparts a yellow color to a wide range of processed foods such as potato chips, corn chips, ice cream and breakfast cereals. Tartrazine is accepted for use in this way by the Food and Drugs Administration. However, it should be noted that in recent years, concern has grown among health professionals and nutritionists over the use of artificial additives in processed foods, including colorants, because of the possibility of chemical change in poorly understood metabolic pathways, which may produce unwanted toxins or carcinogens.

QUESTIONS

1. Explain the phenomena of intramolecular and intermolecular hydrogen bonding, and show how they are responsible for the low solubility of indigo in organic liquids.
2. Explain the modern interpretation of the deep color and semiconducting properties of indigo.
3. Although indigo white is closely related to indigo, explain why indigo white does not have any color.
4. Account fully for the low chemical reactivity of indigo.
5. Explain why adding functional groups to the chromophores of azo dyes may alter their color.
6. Draw the structure of the theoretical, geometric *cis* isomer of indigo.

REFERENCES

C. J. Cooksey. 2001. Tyrian purple and related compounds. *Molecules* 6(9):736–769.

M. Irimia-Vladu et al. 2012. Indigo—A natural pigment for high performance ambipolar organic field effect transistors and circuits. *Advanced Materials* 24(3):375.

O. Meth-Cohn, M. Smith. 1994. What did W. H. Perkin actually make when he oxidised aniline to obtain mauveine? *Journal of the Chemical Society Perkin* 1:5–7.

J. Wouten, A. Verhecken. 1991. High-performance liquid chromatography of blue and purple indigoid natural dyes. *Journal of the Society of Dyers and Colourists* 107:266–269.

SUGGESTED FURTHER READING

J. Balfour-Paul. 1998. *Indigo*. Pages 218–220. British Museum Press.

J. Cannon, M. Cannon. 2007. *Dye Plants and Dyeing*. A & C Black.

Editorial Committee. 1964. *Dye Plants and Dyeing—A Handbook*. Brooklyn Botanical Garden, Brooklyn, New York.

A. Feeser. 2013. *Red, White and Black make Blue: Indigo in the Fabric of Colonial South Carolina Life*. University of Georgia Press.

I. Grae. 1974. *Nature's Colors, Dyes from Plants*. Macmillan Publishers.

A. Kumar Samanta, A. Konar. 2011. Dyeing of textiles with natural dyes. In *Natural Dyes*. E. Perrin Akcakoca Kumbasar, Ed. Pages 29–56. InTech.

S. Robinson. 1969. *A History of Dyed Textiles*. Studio Vista Limited, London.

E. Steingruber. 2004. *Indigo and Indigo Colorants. Ullmann's Encyclopedia of Industrial Chemistry 2004*. Wiley-VCH.

RED DYES FROM HENNA, DYER'S BUGLOSS AND MADDER

Abstract: Many natural and artificial coloring substances, whether they are dyes or pigments, are derivatives of quinones which may be extracted from a range of plants. Quinones are second only to synthetic azo dyes in importance as dye materials.

Alizarin, the red-coloring matter well known to artists, and extracted originally from the Madder plant, was the first natural dye to be synthesized from coal tar.

Natural products chemistry introduces

- Quinones; benzo, naphtha and anthra
- Lawsone, a red-orange dye from Henna—a naphthoquinone
- Alkannin, a red-brown dye from Dyer's Bugloss—a naphthoquinone
- Alizarin, a crimson red dye from Madder root—an anthraquinone
- Parietin, an orange dye from Lichens—an anthraquinone
- Substances in use for millennia in various cultures as hair dye, in body art and tattoos, and in the modern world as cosmetics and food colorants.

Curriculum chemistry involves

- Quinones as building blocks in nature
- Quinones as chromophores
- Applications of quinones as pigments in paints and cosmetics and as dyes in food processing and textiles
- Dyeing textiles
- Colorfastness, resistance to fading by ultraviolet radiation and repeated washing
- An understanding of the theory and use of mordants.

QUINONES AS BUILDING BLOCKS IN NATURE

Quinones (Figure 7.22) are oxidized derivatives of aromatic compounds; benzene, and, with fused rings, naphthalene and anthracene.

However, it is important to note that while quinones are conjugated molecules they are not aromatic. As we have seen with other organic compounds, conjugation of unsaturated groups within a molecule does influence greatly the color chemistry

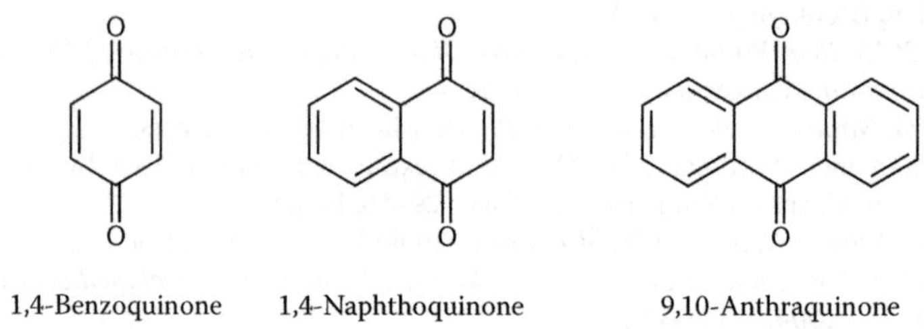

FIGURE 7.22 Quinones.

FIGURE 7.23 **(See color insert.)** Dyeing fabrics in Morocco.

of quinones. The chromophore is the quinone group, which is why the color of many quinones, whether benzoquinones, napthoquinones, or anthraquinones, is red to orange in hue. The concept of chromophores is dealt with in greater depth in the section on "Our World of Green Plants: Human Survival" in this chapter.

Many natural and artificial coloring substances (dyes and pigments) are derivatives of quinones. They are second only to azo dyes in importance as dyestuffs (Figure 7.23). Alizarin (1,2-dihydroxy-9,10-anthraquinone), the red-coloring matter extracted from the Madder plant, was the first natural dye to be synthesized from coal tar. The name, Madder, comes from the term in Arabic, al'isara, meaning "pressed juice."

Naphthoquinones

Lawsone, a Red-Orange Dye from Henna

An extract from the plant Henna (*Lawsonia inermis*) has been used as a colorant for over three thousand years and is often mixed into a paste and used to color skin and hair. Henna has been used as a cosmetic dye on hair and skin and as the pigment in tattoos for millennia, especially on the Indian subcontinent. Lawsone (Figure 7.24) combines chemically with the protein known as keratin in hair and skin resulting in a strong permanent stain that lasts until the skin or hair is shed. Lawsone strongly absorbs ultraviolet radiation, and in aqueous solution is an effective sunscreen.

Even today, henna is used in Ayurvedic medicine for the treatments of rheumatism, insect bites, skin ailments, burns and wounds, to name a few. It is also proven

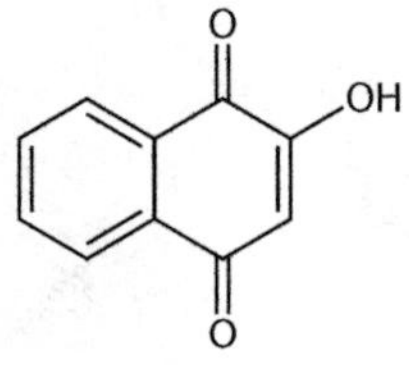

FIGURE 7.24 Lawsone or 2-hydroxy-1,4-naphthoquinone.

to have antifungal and antibacterial properties that are linked to the active component, lawsone—the same chemical that is responsible for its color and properties as a dye.

You can see clearly why the naphthoquinones take their systematic name from the building block, naphthalene.

Alkannin, a Red-Brown Dye from Dyer's Bugloss

Dyer's Burgloss, *Alkanna tinctoria*, is a member of the borage family of plants and provides alkannin (Figure 7.25), a red-brown dye used in food coloring and in the production of cosmetics, especially lipstick. Alkannin is a naphthoquinone with a pentene side chain.

Anthraquinones

The anthraquinones are a common family of naturally occurring substances with yellow, orange and red pigmentation. Anthraquinoids are a class of compounds based upon the anthraquinoid skeleton (see Figure 7.22).

Plant-derived medicinal laxatives containing anthraquinones were presented earlier in the book in the section on "Global Aloe" in Chapter 6. The two key compounds, aloin (Figure 6.20) and emodin (Figure 6.21), have been used as traditional medicines since antiquity.

Yet, many anthraquinones, including emodin, also have a long history of use as dyes. Emodin dyes wool fibers yellow and polyamide fibers red. Fabrics made from these threads show high uptake of this natural anthraquinone, and so emodin has significant potential as a useful alternative to synthetic dyes. There are many natural and artificial coloring substances (dyes and pigments) which are derivatives of quinones. They are second only to azo dyes in importance as dyestuffs.

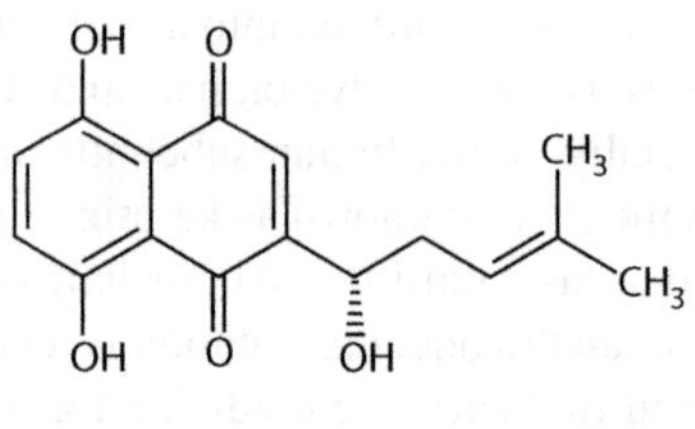

FIGURE 7.25 Alkannin.

Alizarin Red from Madder Root

Early evidence of the use of plant extracts for dyeing textiles comes from Sindh in modern-day Pakistan where cotton dyed with a red substance from the plant, Madder, *Rubia tinctorum*, has been recovered from archaelogical sites dating from 3,000 BC. Alizarin is the crimson red dye involved.

Madder was widely used as a dye in Western Europe too in the late medieval period. In seventeenth century England, alizarin was the red dye used for the uniforms of the New Model Army led by the parliamentarian, Oliver Cromwell, during the English Civil War which began in 1642. The distinctive red color would continue to be worn by soldiers in subsequent centuries giving the English Army, and later the British Army, the nickname, "redcoats."

Alizarin was one of the first natural dyes to be produced synthetically in 1869 which led to the collapse of the market of the dye from its natural source (Figure 7.26).

Alizarin is the main ingredient for the manufacture of pigments known to painters as Rose Madder and Alizarin Crimson. A notable use of alizarin in modern times occurs in biochemical assay to determine quantitatively by colorimetry, calcium and calcium compounds, which are stained red or light purple. Alizarin also continues to be used commercially as a red textile dye.

Parietin from the Lichen, *Xanthoria parietina*

Lichens are plant organisms involving fungi and algae living in symbiosis (see the section on "Our World of Green Plants" in this chapter). Lichens can grow almost everywhere in the world including hot desert climates and snow-packed environments, but favor moist temperate zones where they can be found on rocks, roofs, walls and tree trunks.

Lichens have been used to provide textile dyes, especially for wool, over many, many centuries in northern climes. No mordant is needed to set the color. Many lichens that were used to dye cloth were fermented with ammonia to produce the dye. The natural source of ammonia was urine!

During the Middle Ages, one species of lichen, *Rocella,* was extremely important and known as Royal Purple, and as an extract, was used exclusively by the church and royalty for purple robes.

One commonly found lichen is Crottal or Crottle, *Parmelia saxatilis*, which furnishes a reddish brown dye. It is also known as the skull lichen because it is light gray in color and form.

Parietin is an orange dye extracted from the lichen, *Xanthoria parietina* (Figure 7.27), which is the orange-yellow lichen found in abundance on rocks, walls and tree

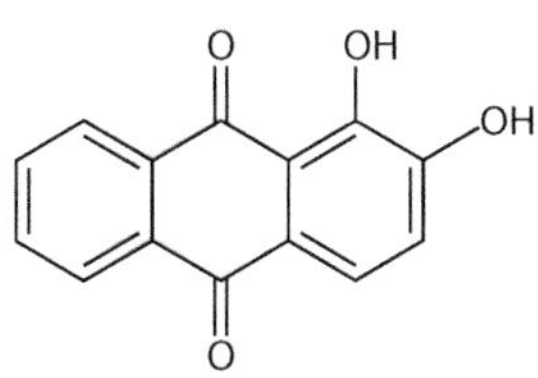

FIGURE 7.26 Alizarin or 1,2 dihydroxy anthraquinone.

FIGURE 7.27 **(See color insert.)** The common lichen, *Xanthoria parietina.* (From http://www.freebigpictures.com)

trunks. The lichen usually grows in exposed positions. As parietin absorbs in the blue end of the spectrum, it protects the lichen from sunlight.

Parietin (Figure 7.28) is the natural dye used for centuries to color both Scottish tartan and the famous cloth woven and dyed in the Outer Hebrides of Scotland called Harris Tweed.

By the early twentieth century, natural dyes had been entirely replaced by organic synthetic dyes, as supplies of these were readily available, and demand for Harris Tweed was growing rapidly worldwide. In the modern world, tartan and Harris Tweed have literally become established as the stuff of modern fashion icons!

Although in modern times lichens are not much used as sources for the dye industry, they do provide an important role in monitoring air pollution. Because most lichens are sensitive to air pollution, the health of the lichen species can be mapped and monitored and used as a proxy measure of air quality especially in industrial zones.

Incidentally, you can see that parietin is a little unusual as a naturally occurring compound in having an ether linkage, a relatively unreactive functional group (see also the section on "Maca from the High Andes in South America" in Chapter 4).

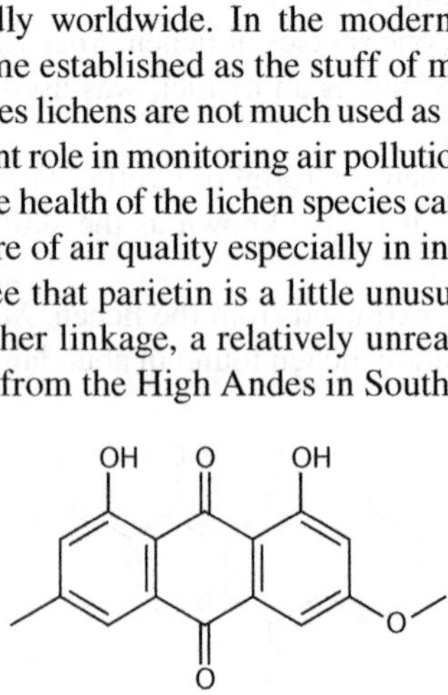

FIGURE 7.28 Parietin.

Textile Dyes: Colorfastness and Mordants

Solubility of Dyes in Water

In order to permeate the fibers of the fabric, dyes need to be soluble in water ideally because water is a cheap and readily available solvent which is straightforward to use. Water is a polar solvent which will readily dissolve polar or ionic solutes due to hydrogen bonding. Sometimes, the molecules of a dye are fully covalent and need to be made polar in order to dissolve the dye in water. This may be done, for instance, by incorporating a sulfate ion in the molecule of a dye to produce a sodium or potassium salt.

Colorfastness

Dyes have to be colorfast to be effective, meaning that over time, the dye should neither fade nor wash out of the fabric easily. For this to be so, physical or chemical interactions have to take place between the dye and the fabric.

The fibers of naturally occurring fabrics, such as cotton and linen, are composed of cellulose which has many hydroxyl groups along its length. Should the dye molecule contain the amine functional group then linkage of the fiber and dye by the electrostatic forces of hydrogen bonding is perfectly feasible. The problem is that such an arrangement does not lead to permanent color dyeing of the textile.

The stronger electrostatic forces of ionic bonding can be taken advantage in some circumstances. A case in point is where the reverse to the above applies, that is, the dye molecule contains an hydroxyl group as part of a carboxylic acid function, while the fiber of the fabric has a complementary alkaline component in the form of an amine group. The acid functional group donates a proton to the alkaline amine, thus creating a distribution of negative and positive centers leading to ionic bonding between the dye and fiber. This situation applies in the dyeing of the natural fabrics of silk and wool and to the artificial fabric, nylon. Colorfastness due to ionic bonding is improved compared with that available through weaker hydrogen bonding.

Of course, the ultimate in colorfastness is achieved through covalent chemical binding of the molecules of dye to the polymers of the fiber. Thus, retention of permanent color by the fabric is the result of chemical reaction between a suitable functional group on the dye molecule and either the hydroxyl group or the amine group typically present in the polymeric molecules of fibers.

Mordants

A mordant is a substance used to affix a dye to a fabric (or to a tissue section in preparation for analysis in the biology laboratory). A particular mordant is chosen specifically for each combination of dye and fabric.

The mordant is usually an aqueous solution of one of a variety of metal salts; those of chromium, aluminum, tin, copper or iron. An example of a material used as a mordant is the mineral deposit, alum, which is a source of chromium, or aluminum ions.

In practice, the fibers of the fabric are first treated with an aqueous solution of the metallic salt and then the dye is introduced. The metal ions are involved in the formation of strong chelates (or coordination complexes), which bind the fiber and the dye.

For background on chelates, see also the section on "Our World of Green Plants: Human Survival" in this chapter. Established interpretation of the formation of chelates is in terms of ligand field theory, which stresses covalent bonding between the ligands and the central ion, and crystal field theory, which emphasizes ionic interactions. These theories account for the perturbing of the outer orbitals occupied by the valence electrons of the central enclosed ion to create quantum steps between energy levels which correspond with the energy available at frequencies in the visible spectrum of light. Hence, absorption of light causes color in many chelates too. Full explanations of ligand field theory and crystal field theory are given in books, for instance, by Murrell et al., and by Barrow, which are mentioned under the subheading below containing suggestions for further reading.

QUESTIONS

1. Alizarin is one of 10 isomers of dihydroxyanthraquinone. How many can you identify?
2. There are only two isomers of benzoquinone. Give their systematic names. Knowing the organic chemistry of the alkene and ketone functional groups, describe the kinds of chemical behavior you would expect of these conjugated molecules.
3. Account for the fact that quinone molecules are strong chromophores, and give examples of the commercial use of these substances as dyes and pigments.
4. Explain differences between dyes which are direct in operation and those dyes which require fixing with a mordant.
5. Lichens are particularly sensitive to the presence of compounds of sulfur. Explain how studies of lichens over time can yield information about the distribution and intensity of air pollution.
6. Certain molecules in organic chemistry, especially those containing the oxygen atom or the nitrogen atom, can act as ligands to form chelates or coordination compounds with the ions of metals, such as magnesium and iron. Explain why the color of copper sulphate changes dramatically from white, as an anhydrous ionic salt, to blue in aqueous solution and then to dark blue when ammonia or an amine, such as methyl amine is added to the solution.
7. What is a mordant? Comment on the fact that bixin, a yellow dye (see the section on "Saffron and Carotenoids: Yellow and Orange Dyes"), does not require the use of a mordant when it is applied in the dyeing of fabrics which are based on cellulose, such as cotton.
8. Phthalo blue is a dark blue pigment used in paints and ink which was developed synthetically in the 1930s. Phthalo blue has the empirical formula, $C_{32}H_{16}N_8Cu$. Explain how the central copper atom forms a coordination compound with phthalocyanin which has four nitrogen atoms each with a lone pair of electrons. Also, explain what is meant by a colloidal suspension of phthalo blue in water, since this is commonly known as ink.

REFERENCES

A. C. Dweck. 2002. Natural ingredients for colouring and styling. *International Journal of Cosmetic Science* 24(5):287–302.

H. Puchtler, S. Meloan, M. Terry. 1969. On the history and mechanism of alizarin red stains for calcium. *The Journal of Histochemistry and Cytochemistry* 17(2):110–124.

SUGGESTED FURTHER READING

G. M. Barrow. 1966. *Physical Chemistry.* McGraw-Hill.

H. S. Bien, J. Stanwitz, K. Windelich. 2005. Anthraquinone dyes and intermediates. *Ullmann's Encyclopaedia of Industrial Chemistry.*

J. N. Murrell, S. F. A. Kettle, J. M. Tedder. 1965. *Valence Theory.* John Wiley.

REVERSIBLE COLORS IN FLOWERS, BERRIES AND FRUIT

Abstract: The brilliant colors of red, purple and blue in many naturally occurring and cultivated ornamental flowering plants, and in their fruit and berries, are due to the production of a class of organic compounds called flavonoids of which the anthocyanins are an important group.

Natural products chemistry

- Flavonoid dyes and pigments
- Anthocyanins.

Curriculum chemistry

- Color in flowers and cross-pollination
- Color in berries and fruit, and food and seed dispersal
- The reversible colors of the acid/base indicators used in chemistry and biology
- Photoperiodism—the significance of phytochrome—a photoreversible pigment.

The Inestimable Value of Color in Flowers, Berries and Fruit

The red, blue and purple colors of many, many naturally occurring and cultivated ornamental flowering plants is due to the production of a class of organic compound called anthocyanins in the flowers and other tissues such as leaves and stems. Particularly clear examples of such flowering plants are pansies (Figure 7.29), violets

FIGURE 7.29 **(See color insert.)** Purple pigmentation due to anthocyanins in the common pansy. (With permission from GNU Free Documentation License.)

FIGURE 7.30 (See color insert.) Strawberries, raspberries, blackberries and blueberries. (Courtesy of http://www.wallpaperup.com)

and snapdragons, while red cabbage and rhubarb are representatives of plants with colored leaves and stems.

Many varieties of insects are attracted to colorful flowers and in many instances there is a specific relationship between the flowering plant and a given insect. The outcome is cross-pollination of the flower and regeneration and dispersal of the species.

Color in berries and fruit (Figure 7.30) is also due to the presence of anthocyanins. While most plants produce anthocyanins, certain plants are particularly rich in them, notably blueberry, cranberry, raspberry, blackberry, blackcurrant, cherry, grapes, red-fleshed peaches and tomatoes.

These strong colors attract herbivores and omnivores from the animal kingdom. As a consequence, the food provided results in the deposition of indigestible seeds, and the dispersal of the plant species whose growth is also conveniently given a boost of fertilizer in the process!

The Beautiful Color of Autumn Leaves

In the autumn, the red color of leaves in deciduous plants, notably trees, is due to the presence of anthocyanins, which according to Archetti et al. are actively produced toward the end of summer. The yellow and orange colors of autumn leaves are the result of the breakdown of chlorophyll which exposes the carotenoids present, xanthophylls (yellow) and carotenes (orange)—see the section on "Saffron and Carotenoids: Yellow and Orange Dyes" in this chapter.

Flavonoids

Several classes of phenolic compounds have been presented in earlier chapters, examples being lignans and lignin from a phenylpropanoid skeleton. This section

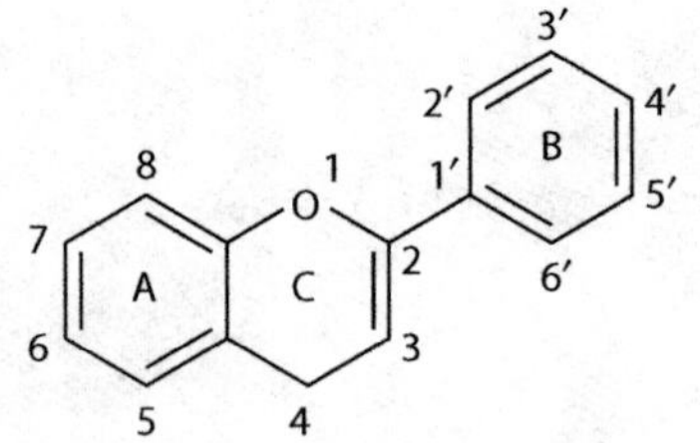

FIGURE 7.31 Common flavonoid carbon skeleton structure.

introduces another important class of phenolic natural products, namely the flavonoids. Flavonoids are the largest class of polyphenols. Chemically, they may be defined as a group of polyphenolic compounds consisting of substances that have two substituted benzene rings connected by the chain of three carbon atoms and an oxygen bridge, as shown in Figure 7.31.

Flavonoids are ubiquitous in plants and products obtained from them such as fruit, vegetables, herbs and beverages, such as tea and red wine. Flavonoids are associated with a broad spectrum of beneficial effects on human health attributable to their antioxidative, antiinflammatory, antimutagenic and anticarcinogenic properties.

The three rings labeled A, B and C are arranged in a fixed pattern. One position in the central ring is occupied by an oxygen atom rather than by a carbon atom. Each atom in the ring is identified by being given a number, running from 1 through 8 for the adjacent rings A and C, while ranging from 1′ through 6′ on the offset ring B.

Flavonoids and Anthocyanins

The flavonoid class of compounds contains the C6–C3–C6 skeleton. On the basis of the degree of oxidation of the 3-C bridge, and whether the bridge is open or closed in a fused ring, flavonoids can then be subdivided into six major groups, among which are the flavonols, flavones, isoflavones and anthocyanins (Figure 7.32).

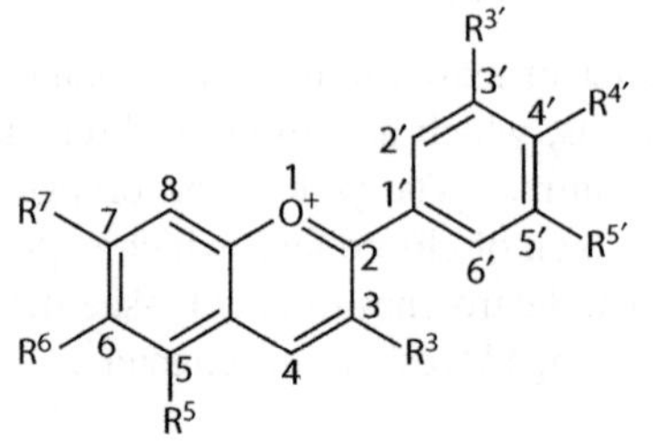

FIGURE 7.32 The generalized molecular structure of anthocyanins where R represents different functional groups.

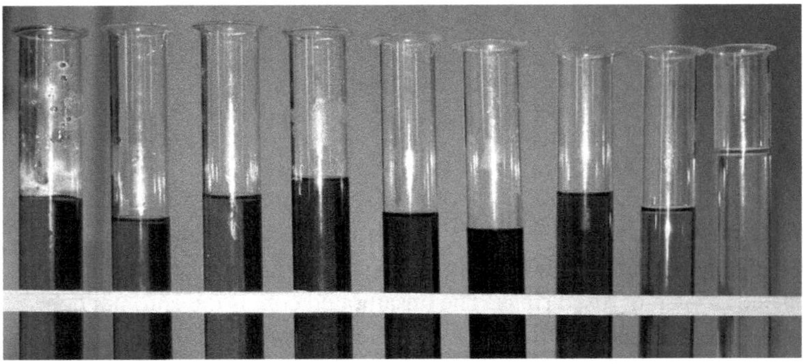

FIGURE 7.33 **(See color insert.)** Anthocyanins from red cabbage change color reversibly dependent on the pH of the solution, from left to right 1, 3, 5, 7, 8, 9, 10, 11 and 13 (With permission from GNU Free Documentation License.)

Anthocyanins: Reversible Dyes

Anthocyanins are chromophores (Figure 7.33). They absorb in the green to yellow parts of the visible spectrum of light dependent upon the pH of the aqueous solution in which the anthocyanin is dissolved, thereby producing the red to purple to blue coloration in light reflected from or transmitted through the tissues of the plant.

Anthocyanins may also act as a sunscreen for the plant as they absorb in a different part of the visible spectrum to chlorophyll.

Anthocyanins are secondary metabolites which find application as a food colorant and are approved for use as such in the European Union, Australia and New Zealand.

Since anthocyanins are water-soluble substances, they can easily be extracted with water from a range of colored tissues in plants such as leaves (red cabbage); flowers (petals of the pelargonium, poppy, rose); berries (blueberries, blackcurrants) and stems (rhubarb). Anthocyanins are red in acidic solutions, purple in neutral conditions, blue in basic solutions and colorless in strongly alkaline solutions.

As these color changes are quite reversible, anthocyanins extracted from plants can provide a crude pH indicator.

Reversible Dyes and Acid/Base Indicators

Litmus

Litmus is a naturally occurring, water-soluble mixture of different dyes extracted from lichens, especially *Roccella tinctoria* (see also the section titled "Our World of Green Plants: Human Survival" in this chapter). Litmus is a naturally occurring pH indicator. The name arises from norse meaning literally "colored moss." The color of litmus changes from red in acidic solution to blue in alkaline solution. The term, litmus test, has entered everyday language as a metaphor for a test that purports to distinguish authoritatively between alternatives.

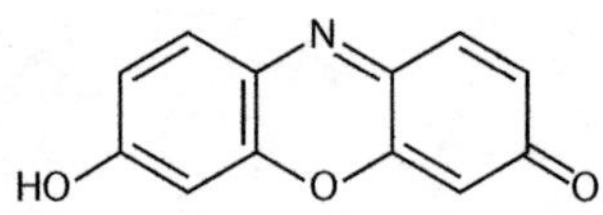

FIGURE 7.34 Phenoxazone.

Litmus compounds belong to a family of substances known as phenoxazones (Figure 7.34), which have complex structures involving a common, unsaturated core involving nitrogen and oxygen in fused rings. The unsaturated core is the chromophore.

As with other indicators, litmus is a weak acid which partially dissociates in water into a hydrogen ion and an anion. The position of the equilibrium is affected when acid or alkali is added to the solution. Acid causes the equilibrium to be displaced toward the undissociated molecule, which is red, while alkali shifts the equilibrium toward the anion as hydrogen ions are removed. The anion is blue. Neutral solution occurs at a pH of around 7, but the purple color is reached rather slowly and indistinctly, which explains why litmus is often used simply to indicate whether a solution of a compound is either acidic or alkaline. The position of the equilibrium is readily reversible caused by the addition or removal of hydrogen ions.

Methyl Red and Methyl Orange

Methyl red is a reversible azo dye, which is made synthetically, as was explained in the section on "Woad and Indigo." Indicators, such as methyl red and methyl orange, are usually weak acids or bases. Methyl red is used in microbiology to identify bacteria which produce acids from the fermentation of glucose, whilst methyl orange is employed in chemistry as an indicator in titrations of weak bases against strong acids. A core structure is common to both substances, which relates to a common reversible color change from red through orange to yellow. The molecular structure of methyl red is shown in Figure 7.35.

Methyl red is a weak acid, which will partially dissociate in water at the functional group of the carboxylic acid. In acidified solution, a hydrogen ion from the strong acid attaches to one of the double-bonded nitrogen atoms, thereby breaking the double bond leaving only a single bond between the neighboring nitrogen atoms. Both the undissociated molecule and the anion are structurally modified and the positive charge from the hydrogen ion is distributed over the structure. As a different chromophore is now present, the solution appears red because the chromophore absorbs in the blue/green area of the visible spectrum.

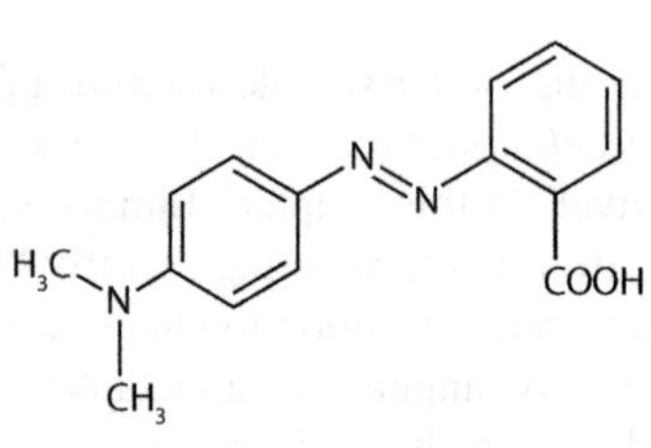

FIGURE 7.35 The molecular structure of methyl red.

The process is reversible. In alkaline solution, the anion appears yellow because it absorbs in the blue area of the visible spectrum. The yellow form of the anion reappears because the nitrogen atom is not protonated and a different chromophore is present with a structure similar to that shown in Figure 7.35.

While the color change is reasonably sharp, methyl red turns orange at a pH of about 4. An equal mixture of the two different anions in methyl orange produces an orange color at about pH 5 rather than at the neutral value of pH 7, so methyl orange is only useful as an indicator in certain circumstances, namely in titrations of weak bases against strong acids.

Phenolphthalein

Phenolphthalein is another indicator used in acid–base titrations. Phenolphthalein does not occur naturally—it is a synthetic compound. Phenolphthalein is also weak acid which dissociates into an H^+ ion and a large anion in solution. Phenolphthalein molecules are colorless in acidic solutions and pink in alkaline solutions when the phenolphthalein anion is predominantly present. When a base is added to the phenolphthalein solution, the molecule/ion equilibrium shifts to the right, as H^+ ions are removed, leading to the release of more anions.

Phytochrome: A Reversible Pigment and a Biological Light Switch: Photoperiodism

The pigment, phytochrome, brings deciduous green plants "back to life" in the Spring in temperate zones of the world, and helps all green plants to set their daily rhythm.

Phytochrome, consists of a bilin molecule—an open, linear chain of four pyrrole building blocks—bonded to a protein. Bilin may be compared with the closed rings containing four cyclic pyrrole molecules found in chlorin and heme covered in the section on "Our World of Green Plants: Human Survival" in this chapter. You will immediately appreciate that phytochrome is not a flavonoid (and therefore, not an anthocyanin), but it does have a very interesting and hugely significant property as a chromophore.

Many flowering plants and deciduous trees make use of phytochrome to assess the relative length of night and day to set their circadian rhythm (or 24-hour cycle), and perhaps, phytochrome is the most important factor (temperature being another) in identifying the changing of the seasons in temperate climes. The circadian rhythm is so named from Latin, *circa* meaning *about* and *dies* the word for a *day*. Thus, a plant can set a period for flowering, and a deciduous tree can grow fresh leaves for the Spring season, and shed leaves in the Autumn season.

Phytochrome is a photoreversible pigment (Figure 7.36). Phytochrome is sensitive to red light, which is readily available from sunlight during the day, absorbing strongly at a wavelength of 650 nm, and thereby producing a blue color. When phytochrome absorbs energy, it changes physically to adopt the form of another conformer (see Glossary) because rotation of groups occurs around the axes of chemical bonds (Figure 7.36). In turn, the conformer is able to absorb strongly at a lower frequency and longer wavelength of 725 nm which is just within the infrared region (sometimes referred to as the near-infrared). When the conformer absorbs, it reverts to

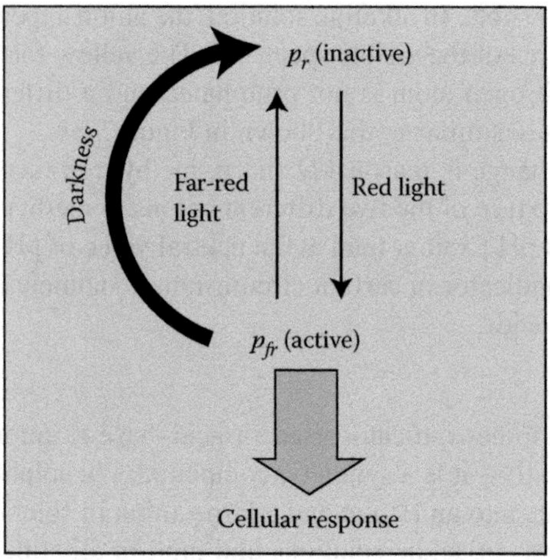

FIGURE 7.36 Photoreversible phytochrome. (Courtesy of http://www.boundless.com)

the ground state once again and appears greenish in hue due to the absorption in the near-infrared. Infrared radiation is present at night when the environment cools and there is no red light from insolation, and so the ground state conformer becomes plentiful in the green plant once again by dawn.

Phytochrome FR is biologically active in the plant, whereas phytochrome R is not.

Many, many green plants use this reversible change from one conformer of phytochrome to another, as both a biological switch and a means of monitoring the length of darkness in relation to the length of daylight. Hence, plants are able to regulate their circadian and seasonal rhythms.

QUESTIONS

1. Compare and contrast the properties of the indicators; litmus, methyl red, methyl orange, thymol blue and phenolphthalein, and explain the most suitable applications for each one in titrations.
2. Explain what is meant by the term end point, or equivalence point, in a titration. Why are indicators not used for titrations of weak acids against weak alkalis?
3. Distinguish between the major subdivisions of the flavonoids; flavanols, flavones, isoflavones and anthocyanins.
4. Resveratrol is a flavonoid with an open bridge structure found in the skin of grapes. Draw its chemical structure.

   Explain why the compound has a chromophore and is responsible for the color of red wine.
5. Give a full account of the significance to green plants, and in turn, to the Earth as a whole, of the reversible photo properties of the pigment, phytochrome.

REFERENCES

M. Archetti et al. 2011. Unravelling the evolution of autumn colors: An interdisciplinary approach. *Trends in Ecology and Evolution* 24(3):166–173.

UK Food Standards Agency. 2011. *Current Approved Additives and Their E Numbers.*

SUGGESTED FURTHER READING

J. Buckingham, V. R. N. Munasinghe. 2015. *Dictionary of Flavonoids.* CRC Press, Taylor & Francis.

J. B. Harborne. 1964. *Biochemistry of Phenolic Compounds.* Academic Press.

L. C. Sage. 1992. *A Pigment of the Imagination: A History of Phytochrome Research.* Academic Press.

W. Vermerris, R. Nicholson. 2006. *Biochemistry of Phenolic Compounds.* Springer.

Glossary

Aliphatic Compounds: The term aliphatic compound refers to any open-chain hydrocarbon such as an alkane, an alkene and an alkyne.

Alkaloids: Alkaloids are a group of naturally occurring chemical compounds, which contain nitrogen atoms within cyclic rings of carbon and hydrogen atoms. The nitrogen atoms are chemically basic. Alkaloids may also contain oxygen and sulfur. Alkaloids are found widely in nature being produced by a large variety of organisms: bacteria, fungi, plants and animals, and are often toxic to other organisms. Although many alkaloids have a bitter taste they have been found to have pharmacological effects in human beings. As a consequence, they are used for medication or as recreational drugs: examples being the local anesthetic and stimulant, cocaine, the stimulants, caffeine and nicotine, the analgesic, morphine, the anticancer compound, vincristine, and the antimalarial drug, quinine.

Antigen: An antigen is a molecule which is foreign or toxic to an organism (a bacterium or a virus), which induces an immune response. An antigen attracts and is bound to a specific antibody prepared by the organism to deal solely with that antigen.

Apoptosis: Apoptosis is the name given to a process of programmed cell death which can occur in multicellular organisms. Cell death follows a controlled path. In general, apoptosis confers advantages during an organism's lifecycle. It is certainly essential for human development; for example, the separation of fingers and toes in a developing human embryo occurs because cells between the digits undergo apoptosis. Apoptosis is also necessary at the start of menstruation.

Research into the effects of apoptosis has increased substantially in recent years because of links with a variety of diseases. Excessive apoptosis causes atrophy as in Alzheimer and Parkinson diseases, whereas an insufficient degree of apoptosis can result in uncontrolled cell proliferation or cancer.

ATP: Adenosine triphosphate (ATP) is an organic substance found in all living cells which is a source of energy and acts rather like an enzyme. ATP transports chemical energy within a cell to allow metabolism to occur. It is important to note that metabolic processes which utilize ATP later restore it. ATP is therefore continuously recycled in organisms.

Chelate: A chelate is a compound containing a ligand bonded to a central metallic ion at one or more points. The International Union of Pure and Applied Chemistry (IUPAC) define chelates as compounds, which involve the presence of two or more separate coordinate bonds between a polydentate (multiple bonded) ligand and a single central atom or ion. See also the definition of a ligand.

Chloroplast: The chloroplast is the place within the cell of a green plant where photosynthesis occurs.

Chromophore: A chromophore is the part of a molecule responsible for its color. The color arises when a molecule absorbs certain wavelengths of visible light and transmits or reflects others. The chromophore is a region in the molecule where the energy difference between two different molecular orbitals falls within the range of the visible spectrum. Visible light that hits the chromophore can thus be absorbed by exciting an electron from its ground state into an excited state.

Conformer: Conformational isomerism occurs through rotation about a single bond in a molecule. It differs from structural stereoisomerism where interconversion would require the breaking and reforming of chemical bonds.

Conjugation: In chemistry, conjugation occurs in compounds with alternating single and multiple bonds, which allows p-orbitals to interconnect and delocalize electrons thereby increasing stability in the molecule. A conjugated compound may be linear, cyclic or mixed.

Critical Point: See the entry for supercritical fluids.

Dalton: A Dalton is a physical constant accepted within the international SI system, which has been named after the famous British chemist, John Dalton, who first proposed its use. The Dalton (represented as u or Da) is a relative unit of mass being based upon one twelfth of the mass of a carbon nucleus. The actual mass of 1u or Da is $1.6605389217 \times 10^{-27}$ kg.

Diastereomers and Enantiomers: Diastereomers (also known as diastereoisomers) arise when two or more stereoisomers of a compound exist, which have different configurations at one of the equivalent stereocenters and which are not mirror images of each other. Stereoisomers, which are mirror images and may be superimposed on one another will rotate the plane of polarized light and are known as enantiomers.

DNA: Deoxyribonucleic acid is a very long polymer made up of monomers called nucleotides. Nucleotides are in themselves complex molecules but along a DNA polymer there are only four kinds of nucleotides each one containing a different base; adenosine (A), cytosine (C), guanine (G) and thymine (T). These four different nucleotides can occur in any sequence along the DNA chain. It is the order which contains the information representing the unique genetic code of an organism.

Electrophile: An electrophile, such as a proton, is an electron acceptor which is attracted to electron-rich parts of molecules and often undergoes a chemical reaction there.

Elimination Reactions: In an elimination reaction, a small group of atoms or a small molecule breaks away from a larger molecule and is not replaced.

Enzyme: An enzyme is a biological catalyst, which speeds up or facilitates chemical reactions in organisms.

Enzyme Inhibitor: An enzyme inhibitor is a molecule that binds to an enzyme and decreases its activity.

Functional Group: In organic chemistry, a functional group is a specific group of atoms or bonds within a molecule, which is responsible for characteristic chemical reactions. However, the relative reactivity of a functional group may be modified by other functional groups nearby and by the structure of the molecule.

Hodgkin's Disease: Hodgkin's disease appears as a tumor in the glands known as lymph nodes found in the neck, groin, armpits and chest. The lymphatic system is a network of vessels which runs throughout the body carrying the colorless fluid, lymph, which transports the white blood cells so important in maintaining immunity from infection. Abnormal white blood cells are produced by the lymph nodes and are a characteristic of Hodgkin's disease.

Hormones: A hormone is a regulatory substance produced in an organism and transported in tissue fluids such as blood or sap to stimulate specific cells or tissues into action.

Hydrogen Bonding: Covalent bonds consisting of a hydrogen atom bonded to an atom of a strongly electronegative element (such as oxygen, nitrogen or a halogen) are highly polar. These polar bonds give rise to permanent dipoles which result in electrostatic attraction between neighboring molecules—known as hydrogen bonding.

Hydrophilic: Meaning water "loving."

Hydrophobic: Meaning water "hating."

Lipids: Lipids are fatty acids which are insoluble in water yet soluble in many organic solvents. The class includes natural oils, waxes and steroids.

Ligand: Compounds exist in which a central metal ion coordinates with electron-rich groups called ligands. The aquated ion of copper is an example, $(Cu(H_2O)_6)^{2+}$, which gives copper sulfate solution its deep blue color—anhydrous copper sulfate being a white powder.

Metabolites: Metabolites are intermediates in metabolic processes in nature and are usually small molecules.

Miscible: The term miscible applies when two or more liquids, such as water and ethanol, are able to dissolve into one another in any proportion without separating.

Mitochondria: In mitochondria within cells, nutrients are broken down to release energy. Enzymes in the mitochondria act in aqueous solution to oxidize molecules of food. The process is known as cellular respiration.

Mitosis: In mitosis, a cell divides to create two identical copies of the original cell.

Mole: One mole is defined as the molecular weight or atomic weight of a substance expressed in grams based on the standard of 12 g of carbon 12.

Nucleophile: A nucleophile is an electron-rich species such as an anion, which undergoes chemical interaction with a molecule by being attracted to atoms in dipolar bonds, which have a positive charge.

Primary Metabolites: Primary metabolites are needed for the growth, development and reproduction of an organism.

Phototropism: Phototropism is the orientation of a plant or other organism in response to light either toward the source of light (positive phototropism) or away from it (negative phototropism).

Redox Reactions: Redox reactions are very important in living organisms. In a redox reaction, electrons are transferred between species, be they atoms, molecules, ions, free radicals, etc. In a reduction reaction, a species gains an electron(s) whereas in an oxidation reaction an electron(s) is lost. In a redox reaction, one species gains an electron (an acceptor) at the same time as another species loses an electron (a donor)—hence the term. An oxidizing agent accepts an electron and is reduced in the process while a reducing agent gives an electron and is oxidized.

RNA: Ribonucleic acid is present in all living cells and in many viruses. RNA consists of a long, usually single-stranded chain of alternating phosphate and ribose units, with one of the bases adenine, guanine, cytosine, or uracil bonded to each ribose molecule. Unlike DNA, RNA is more often found in nature as a single-strand folded onto itself, rather than a paired double strand. Cellular organisms use RNA to convey the genetic information (using the letters G, A, U and C to denote the nitrogen bases guanine, adenine, uracil and cytosine), which directs synthesis of specific proteins. Many viruses encode their genetic information using an RNA genome.

Secondary Metabolites: Secondary metabolites in plants are chemicals, which, thus far, have no identified role in growth, photosynthesis, reproduction or any other "primary" function. These chemicals are extremely diverse, indeed, many thousands are known. They may be classified on the basis of chemical structure into three main groups: the terpenes (composed almost entirely of carbon and hydrogen), phenol-related compounds (made from simple sugars, containing benzene rings, hydrogen and oxygen), and molecules containing nitrogen. The fact that many secondary metabolites have a specific negative impact upon other organisms, such as grazing animals, insects or bacteria, has led to the hypothesis that secondary metabolites have evolved in plants directly because of their protective value.

Steric Hindrance: Steric hindrance occurs when the large size of a particular group within a molecule prevents chemical reactions, which are observed in related molecules without that group. Although steric hindrance is sometimes a problem, it can also be exploited to change the pattern of the chemistry of a molecule by inhibiting unwanted side reactions (known as steric protection) or by leading to a preference for one course of stereochemical reaction over another (as in diastereoselectivity).

Sublimation: Sublimation is said to occur when a substance passes directly from the solid to the vapor phase without going through the intermediate liquid phase (and vice versa).

Supercritical Fluids: Any substance, provided it does not break down chemically and physically, can become a supercritical fluid at a temperature and pressure above its critical point where distinct liquid and gas phases do not exist. Supercritical fluids have properties between those of a gas and a liquid.

As there is no liquid/gas phase boundary, there is no surface tension in a supercritical fluid. By changing the pressure and temperature of the fluid, its properties can be "tuned" to be more like a liquid or more a gas. The solubility of a supercritical fluid tends to increase with density of the fluid (at constant temperature). Since density increases with pressure, solubility tends to increase with pressure—a property which can be exploited in industrial processes.

Tincture: A tincture is an old-fashioned medical term for a solution of an organic compound in ethanol.

Van Der Waals Forces: Van Der Waals forces are forces of attraction which exist between all atoms and all molecules.

They arise from electrostatic attraction between temporary dipoles and induced dipoles caused by the movement of electrons within atoms and molecules. Consequently, Van Der Waals forces are much weaker than any other type of molecular interaction. They are only significant at all in atoms and molecules, which have no other form of intermolecular attraction, examples being nonpolar molecules and the noble gases, Group 0, in the periodic table of elements.

Winchester Bottle or Boston Round: A large glass storage bottle for liquids, sometimes tinted amber to absorb light, used in the chemical industry and in chemical laboratories.

Index

A

Absorption bands, 199–200
Acetaminophen, 153–154
Acetylation, 28–29, 136, 137
Acetylcholine, 128, 129
Acetylsalicylic acid, *see* Aspirin
Acid–base extraction, 50, 125, 126, 147
  indole and, 125
  isolation of cocaine, 147
Acid rain, 99
Addition reaction, 22, 160–161
Adenosine triphosphate, 85
ADP, 6
Aegilops, 67, 68
Aegilops searsii, 71
Aegilops speltoides, 71
Age-related macular degeneration (AMD), 209
Aglycone, 184
Agricultural production, enhancing, 78–80
Air pollution and fossil fuels, 99
Ajoene, 97
Alanine, 75
Alcohols, 40–41, 64, 68–69, 107
  esters formed from, 40–41
  phenols compared with, 107
Aldehyde, 86–87, 160
Aldose, 86
Algae, 94–95, 192
Alizarin, 220, 221, 223
Alkaloids, 49–50, 123, 124, 125–126, 134, 136, 145, 146–147, 149, 151
Alkanals, 160
Alkanes, 19, 20, 40
Alkanna tinctoria, 222
Alkannin, 220, 222
Alkanones, 160
Alkenes, 55, 143
Alkynes, 23
Allium sativum, *see* Garlic
Allopathic medicine, 2, 11
Allyl functional group, 204, 209–210
Allylic position, 209
Allyl methyl sulfide, 97
Aloe vera, 180–181
  anthraquinones found in, 183–184
  chromatography, 185
  galactomannans, 181–182
  gas chromatography, 186
  gas–liquid chromatography, 186
  high-pressure liquid chromatography, 186
  hydraulic fracturing, 182, 183
  ion-exchange chromatography, 186
  polysaccharides and, 181
  skin treatment and, 181
  thin-layer and column chromatography, 185
Aloin, 181, 183–184, 222
α-boswellic acid, 170
α-linolenic acid, 60
Alum, 225
Amazon rainforest tribes, 3–4
Amides, 118, 124, 153–154
Amines, 118–119, 151, 216
  primary, 151, 186
  secondary, 147, 151, 186
  tertiary, 118, 151, 186
Amino acids, 61, 68, 72, 73, 76, 77, 118
  essential, 61, 74–75, 124
  non-essential, 74–75
  protein and, 74–75
  sulfur and, 97–98
Ammonia, 118, 151, 153, 195, 223
Ammonium thioglycolate, 98
Aniline, 33, 216–217
Animalia, 191
Anthocyanins, 228, 229
  flavonoids and, 230–231
  reversible dyes, 231
Anthraquinones, 183–185, 220, 221, 222–224
  alizarin, 220, 223
  parietin, 220, 223–224
Anthropology, 3
Antigen, 40
Apoptosis, 107
Arginine, 75
Aromatic amines, 151, 152, 154, 216
Aromatic compounds, 24, 33, 125, 151–152
Artemisia annua, 34, 35
Artemisinin, 31, 34, 35, 36, 44
  molecular structure of, 35
  peroxide link and peroxide bridge in, 36
Asafetida, 14
Ascorbic acid, *see* Vitamin C
Asparagine, 75
Aspartate, 75
Aspirin, 26, 28–29, 136
  carboxylic acids, 26–27
  phenol and phenolic compounds, 27–28
  salicin and salicyclic acid, 26
Atmospheric pollution, 23, 99, 161–163, 200–201
ATP, 85, 93–94

Ayurvedic medicine, 11–12, 166, 167, 181, 221
Azo dyes, 197, 217–219, 232

B

Badische Anilin und Soda Fabrik (BASF), 217
Basil, 55
Bayer, 26, 29, 137
Beer brewing, 87–88
Beer, Charles, 48
Bees wax, 64
Benzene, 22–23, 197
  halogenation of, 22
  Kekule's theory of, 23–24
  molecular structure of, 23–24
Benzenediazonium chloride, 218
Berries, 229
Berzelius, 207
β-carotene, 196, 197, 207, 208, 210
Beverages, 101–129
  cocoa, 110–114
  coffee, 115–120
  maca, 122–129
  tea, 102–108
Biblical resins, 165
Bilin, 233
Biodiesel, 64–65, 94, 178
Bioethanol, 94–95
Biofuel, 70, 94
Bioplastic, 88
Birth control pill, 14–15, 18
Bixin, 206, 226
Black tea, 104, 105
Boswellia serrata, 165, 167
Boswellic acids, 167–168, 170
Botany, 3
British East India Company, 102, 103
Building block, 6, 42, 47, 50, 53, 124, 151, 176, 177, 193, 213, 220, 250
B vitamins, 42, 43

C

Cacao, 110, 111
Caffeine, 117, 118–120
  coffee and, 117
  resonant molecular structures of, 118
  structure of, 117
  ways to remove coffee from, 119–120
Camellia sinensis, 93, 102, 104
Camphor, 158
  carbonyl group and nucleophilic addition reactions, 160–161
  classification and structure of, 158
  diverse uses of, 158–159
  extraction of, by steam distillation, 159–160
  infrared radiation, absorption of, 161–163
  infrared spectroscopy, 161
  structure of, 158
Camphor oil, 160
Cancer, green tea helps against, 16
Cannabinoids, 140, 141
Cannabis, 140–141
  cannabinoids, 140–141
  condensation reaction, 143
  terpenes, 141–143
Cannabis indica, 140
Cannabis sativa, 140
Carbohydrates, 71–72, 82–89, 94
  chemical properties of, 86–87
  classification of, 82
  commercial uses of, 87–89
  disaccharides, 83–84
  glycoside link, 84–85
  groups of, 83
  human nutrition and, 87
  monosaccharides, 83, 86–87
  oligosaccharides, 85
  polysaccharide, 83, 85–86, 87
  saccharides and, 82–83
Carbolic acid, *see* Phenols
Carbon accounting, 201–202
Carbon neutrality, 201–202
Carbon–oxygen bonds, 34–36, 40, 99
  double, 35, 36, 126
  single, 34–35, 126
Carbonyl group, 49, 153, 160–161
Carboxylic acid, 26–27, 28, 41, 42, 125, 177, 178–179
Cardiac health, cocoa and, 112
Carnuba wax, 64
Carotenes, 195, 206, 207, 208
  β-carotene, 207
  isolation and extraction, 208
  lycopene, 207–208
Carotenoids, 97, 141, 204, 205–207, 208, 229
Catechins, 105, 106–107
Catharanthus rosea, *see* Periwinkle
Cellophane, 88
Cellulose, 69, 85–86, 87, 88–89, 181, 215
Cellulose nitrate, 89
Cereals, 66
Chelates, 225–226
Chenopodium quinoa, see Quinoa
Chia, *see Salvia hispanica*
China
  opium in, 133
  tea in, 103
Chirality, 55–56, 76
  isomers and, 55–56
  on life systems on earth, 76
Chlorella, 192
Chlorin, 193, 194, 195
Chlorobenzene, 22

Chlorophyll, 189, 193–195, 205
Chloroplasts, 193
Chloroquinine, 31
Chocolate, 110–112, 114; *see also* Cocoa
Cholesterol, 63, 105
Chondodendum tomentosum, 128
Chromatography, 50, 185–186
  gas, 186
  gas-liquid, 186
  high pressure liquid, 186
  ion-exchange, 186
  thin-layer and column, 185
Chromophores, 195–197, 210, 214, 221, 232, 233
Chrysoidine, 218
Cinchona, 31
Cinnamomum camphora, 158
Cis-trans isomerism, 55, 106
Climate change, 178, 200–202
Coca, 145, 146
Cocaine, 145–146
  chemical properties of, 146–147
  isolation of, 147
  legitimate applications of, 147–148
Cocoa, 110; *see also* Beverages
  Aztecs and, 110–111
  cardiac health and, 112
  chemical constituents of, 113–114
  consumption, 110
  diabetes and, 112–113
  liquor, 110
  origin of, 110
Codeine, 136, 137, 138
Coffee, 115; *see also* Beverages
  from Arab world, 115–116
  caffeine and, 117, 119–120
  cultivation in Ethiopia, 116
  early use of, 116–117
  history of, 115–116
  slavery and, 116
Colloidal suspension, 176
Colloids, 176
Color, 5, 215, 217
  of autumn leaves, 229
  chlorophyll and, 195
  chromophores and, 195–197
  in flower, berries and fruit, 228–229
Commiphora myrrha, 165
Comstock Act of 1873, 17
Condensation reaction, 63, 73–74, 143
Conformer, 233–234
Conjugation, 199, 220
Contraception, 13–14, 18; *see also* Birth control pill
Controlled Substances Act 1970, 146
Convinca, 47
Cool-season cereals, 69–70
Coordination complexes, 195
Coordination compounds, 195
Cordyceps sinensis, 90
  bioethanol, industrial production of, 94–95
  for enhancing energy, 92
  fermentation, 93–94
  health benefits of, 90–93
  life cycle of, 90
Cosmetics, 175, 181, 221, 222
Coupling reaction, 57, 69, 218
Covalent bonding, 35, 153, 195, 196, 200, 226
Cracking of crude oil, 20–21
Crocin, 205, 206, 208
Crocus cartwrightianus, 205
Crocus satires, 204
Crottal, 223
Crude oil, 20–21, 176
Cuba, slave trade in, 116
Curare, 3, 126, 128–129
Cyamopsis tetragonoloba, 181
Cyclic aromatic amines, 118–119
Cyclic compounds, 24, 151, 152
Cyclic esters, 42, 44
Cyclic terpenes, 168–170
Cycloalkanes, 13, 19–20
Cyclohexane, 20, 22
Cysteine, 75, 97, 98

D

Dahlia tubers, 85
Dalton, 160
Daucus carota, 14
Decaffeination, 50, 119–120
Dehydration, 41, 42, 142–143
Delocalization of electrons, 24, 28, 119, 152, 195, 215
Deoxyribose, 87
Diabetes, 47, 48
  cocoa and, 112–113
  periwinkle for treating, 47
Diacetylmorphine, *see* Heroin
Diallyldisulfide, 97
Diamond, Jared, 66
Diastereoisomers, 106
Diastereoselectivity, 171
Dicarboxylic acids, 27
Diethyl ether, 41
Digitalis, 38–40
Digoxigenin, 38–40, 42
Dimers, 68
Diols, 144
Dioscorea barbasco, 15
Dioscorea Mexicana, 15, 16
Dioscarea villosa, 15
Diosgenin, 15–16
Disaccharides, 83–84
Djerassi, Carl, 16
DNA, 87, 113

Dreser, Heinrich, 137
Dyer's Burgloss, *see Alkanna tinctoria*
Dyes, 190; *see also individual dyes*
allyl functional group, 209–210
anthraquinones, 222–225
azo dyes, 217–219
carotenoids, 207–208
indigo, 211–217
naphthoquinones, 221–222
quinones, 220–221
reversible colors in flowers, berries and fruit, 228–231
reversible dyes and acid/base indicators, 231–234
saffron, 204–207
textile dyes, 225–226
vitamin A, retinol, retinal, 210–211
woad, 211
xanthophylls, 208–209

E

Einkorn wheat, *see T. monococcum*
Electromagnetic radiation, molecular interaction with, 197–198
Electromagnetic spectrum, 198
Electrophilic substitution, 24, 28, 108, 218
indole and, 124–125
reactions of phenol, 107
Eli Lily Company, 47, 48–49
Elimination reaction, 142–143
Eluate, 185
Emmer wheat, *see T. dicoccum*
Emodin, 184–185, 222
Enantiomers, 56, 146, 154, 160
Endorphins, 138
Enovid, 17–18
Enzyme inhibitor, 75
Enzymes, 72–73, 76, 85, 88, 97
Epicatechin, 106, 107, 114
Ergometer, 92
Ergotamine, 127
Erythroxylon coca, 145
Essential amino acids, 74–75
Essential oils, 141, 157, 167, 168, 170, 171, 175–176
Esterification, 86, 89
Esters, 40–42, 44, 63–64, 176, 177
formed from alcohols and acids , 40–41
hydrolysis of, 177
properties and uses of, 41–42
Estrogen, 14–15
Ethanol, 40, 41, 57–58, 88, 93, 94–95, 119, 185
Ethene, 41
Ethers, 35–36, 86, 126, 154, 214
Ethiopia, 116
Ethnobotany, 3
Eugenol, 78
Euphorics, 131
cannabis and marijuana, 140–143
coca and cocaine, 145–148
morphine, 132–138
tobacco, 149–155
European lavender, 173
colloids and hydrosols, 176
distribution, 174
historical note, 174
hydrolysis of esters, 177
hydrosol from, 175–176
soap, 177–179
value of, 175
vegetable oils and fats, 176–177
Exotic fragrances, 157
Explosives, 89

F

Fats, 63–64, 176–177
Fat-soluble vitamins, 42–43
Fatty acids, *see* Lipids
Feedstock, 41, 88, 217
Fermentation, 94–95, 105, 111
Flavonoids, 113–114, 141, 228, 229–230, 234
anthocyanins and, 230
definition, 229
Flavonol, 106, 107, 112
Flowering plants, 192
Folk medicine, 2, 101
Food and Drug Administration (FDA), 159
Foods and grains; *see also individual foods and grains*
chia and quinoa, 60–65
Cordyceps sinensis, 90–95
garlic and pungent smells, 96–99
rice, 81–89
wheat, 66–80
Fossil fuels and air pollution, 99, 201
Foxglove, 38–40
Fracking, 182–183
Fractional distillation, 20–21, 50, 175–176
Frankincense, 165, 166, 167, 168, 171, 216
Free radicals, 35, 36, 107, 113
Friedel-Crafts reaction, 22
"Frothy jade", *see* Matcha tea
Fructose, 82, 85, 86, 87
Functional group, 6–7, 49, 57, 62, 76, 86, 89, 126, 143, 161, 186, 209–210
Fungi, 90, 191

G

Gaja-bhaksha, *see* Frankincense
Galactomannan, 86, 181–182, 250
γ-nonalactone, 39, 44

Garlic, 96–97
amino acids, 97–98
fossil fuels and air pollution, 99
organosulfur compounds, 97
Gas chromatography, 186
Gas–liquid chromatography, 186
Genetic modification of crops, 66, 78–80
Geometric isomerism, 55, 213–214
Gibberellins, 44
Gliadin, 61–62
Global warming, 88, 161–163, 183, 200, 201
Globin, 195
Glucose, 26, 83, 84, 86, 87, 88, 89, 112, 181, 182, 193, 214, 250
Glutamic acid, 75
Glutamine, 75
Gluten sensitivity, 61–62, 71
Glycerol, 63, 173, 176–177
Glycine, 75
Glycoside, 26, 38, 83–85, 180, 181, 183–184, 214
Glycoside link, 84–85, 181
Glyphosate, 79
Gramineae, 66
Greenhouse effect, 158, 161–163
Greenhouse gases, 162–163, 178, 200
Kyoto Protocol, 162–163
Montreal Protocol, 163
Green issues, 66
Green plants
algae, 192
chlorophyll, 193–195
chromophores, 195–197
color and chlorophyll, 195
electromagnetic radiation, molecular interaction with, 197–198
flowering plants and conifers, 192
lichens, 192
limiting climate change, 200–202
photosynthesis, 193
ultraviolet absorption spectroscopy, 198–200
Green tea, 104
benefits of, 105
catechin in, 106–107
helps against cancer, 106
polyphenols in, 107
Guar gum, 182–183
Gun cotton, 89
Guns, Germs and Steel (Diamond), 66

H

Hair, 36, 98, 221
Halo alkanes, 6
The Handbook of Prescriptions for Emergencies, 34
Harrison Narcotics Tax Act of 1914, 138
Hashish, 140
Henna, 183, 185, 195, 196, 220, 221–222
Heroin, 132, 133–134, 138
chemical composition and properties of, 136–137
morphine conversion to, 136, 137
Heterocyclic aromatic compounds, 7, 24, 119, 125, 149, 151–153
pyridine, 152
pyrrole, 152
thiophen, 152–153
Heteropolysaccharides, 83, 86
High-pressure liquid chromatography, 186
Histidine, 75
Hodgkin, Thomas, 47
Hoechst, 217
Hoffmann, Felix, 29
Hood, Robin, 215, 216
Hormones, 13, 14–15, 16, 44
Human health, 43, 63, 79, 155, 204, 210–211
Humble potato, 13
benzene and aromatic compounds, 22–24
birth control pill, 14–15
contraception, 13–14
crude oil and fractional distillation, 20–21
global emergence of yam, 18–19
hydrocarbons, 19–20
isolation of diosgenin from Mexican yam, 15–16
norethindrone, 16–17
oral contraception, 18
progesterone to prevent ovulation, 17
social acceptance and medical approval, 17–18
steroids, 13
Hydraulic fracturing, 182, 183
Hydrocarbons, 19–20, 23, 52, 141, 162, 163
classification of, 19
examples of, 20
saturated, 19–20
Hydrogen bonding, 27, 36, 72, 178, 199, 215, 225
Hydrolysis, 66, 74, 83, 177, 184, 214
Hydrophilic, 178
Hydrophobic, 175, 178
Hydrosol
colloids and, 176
from lavender, 175–176
Hydroxyl group, 26, 27–29, 40, 106, 107, 136, 137, 209, 225

I

Immunohistochemical staining (IHS), 40
Incense, 165
Indican, 214
Indicators, 231–233
litmus, 231–232
methyl orange, 232–233

Indicators (*Continued*)
methyl red, 218, 232–233
phenolphthalein, 233
Indigo, 197, 202–203, 212–217, 252
through ages, 212
Indigo
azo dyes, 217–219
dyeing textiles and, 215–217
extraction of, 214
historical note on, 212–213
from natural sources, 214
physical and chemical properties of, 214–215
from plants, 213–214
Indigofera tinctoria, 213
Indigotin, 213
Indigo white, *see* Leuco indigo
Indole, 47, 49–50, 122, 124–125, 213
bases and, 125
chemical properties of, 124–125
electrophilic substitution, 124–125
Indole alkaloids, 126
purification, by acid–base extraction, 125–126
types of, 126
Indole terpenoid alkaloids, 126
Indoxyl, 214
Infrared radiation, 161–163, 200, 201, 233–234
Infrared spectroscopy, 161
Inulin, 85
Ion-exchange chromatography, 180, 186
Isatis tinctoria, 5, 190, 197, 212, 213
Isoleucine, 75
Isomerism, 7, 40, 55, 57, 106, 213–214
chirality and, 55–56
definition, 54
stereochemistry and, 54–55
Isoprene, 53–54, 126, 142, 158, 170, 206–207, 210
Isoprene rule, 140, 142, 170

J

Johnson, Caleb, 178

K

Kekule, 23, 24, 217
Keratin, 98, 221
Ketohexose, *see* Fructose
King, H. J., 128
Krebs cycle, 5, 6
Kuna Indians, 112
Kyoto Protocol, 162–163

L

Lactic acid, 5, 42, 93, 134
Lactones, 38, 42–43, 44
as building blocks in nature, 42
uses of, 44
Lactose, 83, 84, 85
Laudanum, 133
Lavandula angustifolia, 174
Lavender, 173, 174, 175–176
Lavoisier, Antoine, 193
Lawsone, 220, 221–222
Lawsonia inermis, *see* Henna
Lepidium meyenii, *see* Maca
Leuco indigo, 215
Levi Strauss, 215
Lichens, 191, 192, 223, 224, 231
Ligands, 129, 195, 226
Light sources, frequency ranges of, 198
Lignans, 66, 68–69, 229
Lignin, 66, 68, 69, 229
Limonene, 141, 142, 143, 170, 171
pinene, 54, 158
squalene, 5, 142
Linalool, 175, 176
Linalyl acetate, 175, 176, 177
"Lincoln Green", 215–216
Lipids, 60, 62, 63, 64, 94, 177, 178
Litmus, 231–232
Lone pair electrons, 151, 195
Lutein, 206, 208–209
Lycopene, 97, 207–208
Lysergic acid diethylamide (LSD), 127, 133
Lysine, 61, 75

M

Maca, 101, 122; *see also* Beverages
as beverage and food, 122–123
chemical composition of, 123
indole, 124–125
indole alkaloids, 126–129
medicinal value of, 123
purification of indole alkaloids, 125–126
Madagascan periwinkle, *see* Periwinkle
Madder, 220, 221, 223
The Magi, 165–166
Malaria
artemisinin, 34
carbon–oxygen single bonds, 34–36
cinchona, 31
drugs for, 31
oxygen–oxygen single bonds, 34–36
peroxide link and peroxide bridge in artemisinin, 36
quinine, 32–34
symptoms of, 31
Maltose, 88
Marker, Russell, 15, 16
Matcha tea, 104
Mauveine, 216, 217

Mayan Empire, 150
Medicinal plants, 1–2, 4, 11, 12, 47, 49; *see also* Ayurvedic medicine
Mendeleev, 99
Metabolites, 5, 43, 52, 67, 68, 231
Methionine, 75, 97–98
Methyl orange, 232–233
Methyl red, 218, 232–233
Mexican yam, 15
  global emergence of, 18–19
  isolation of diosgenin from, 15–16
Millennium Seed Bank, 163
Miramontes, Luis, 16
Mole, 20
Molecular spectroscopy, 198
Monera, 191
Monomer, 143
Monosaccharides, 82–83, 85, 86, 87, 88, 181
Monoterpene, 54
Montreal Protocol, 163
Mordants, 220, 223, 225–226
Morphine, 4, 132–138
  case study, 132–133
  chemical structure of, 135
  conversion to codeine, 136
  conversion to heroin, 136, 137
  medicinal uses of, 138
  production of, 134–135
  purification and properties of, 135–136
Myrrh, 165–166, 168–171
Myrtenol, 216

N

NAD+ , 93–94
NADH, 93–94
Naphthalene, 197, 217, 220, 222
Naphthoquinone, 221
  alkannin, 222
  lawsone, 220, 221–222
National Cancer Institute (NCI), 52
National Organic Standards Board (NOSB), 154
Natural products, chemistry arising from, 8–10
Neonicatinoids, 154
Nicotiana rustica, 149
Nicotiana tabacum, 149
Nicotine, 131, 149, 151, 154–155
  mood-altering effects of, 155
  properties of, 154
  uses of, 154–155
Nicotinamide adenine dinucleotide (NAD), 94
Nitriles, 151, 160, 161
Noble, Robert, 48
Nomenclature, 7, 26, 42, 165, 170
Nonalcoholic beverages, 101
Nonessential amino acids, 74, 75
Norethandrolone, 17
Norethindrone, 16–17
Norethynodrel, 17
Nuclear magnetic resonance (NMR) spectroscopy, 52, 56–58, 75–76, 199
Nucleophiles, 22, 24, 158, 218
Nucleophilic addition reactions, 160–161
Nucleophilic substitution, 22, 102

O

Ocimene, 55
Oils, 63, 64–65, 78, 141, 157, 165, 171, 173, 175–177
Oleic acid, 62, 111, 123
Olibanum, *see* Frankincense
Oligosaccharides, 82, 85
Omega-3 fatty acids, 60
Opium, 131, 133, 134–136, 138
  in China, 133
  harvesting of, 134–135
  history of, 133–134
  tea and, 103
Oral contraception, 13, 18
Organic chemistry, 5–7, 8–10, 56, 82, 86, 143, 198–199
Organic farming, 154
Organosulfur compounds, 96, 97, 99, 163
Ovulation, progesterone to prevent, 17
Oxygen–oxygen single bonds, 31, 34–36
Ozone, 23, 113, 163
Ozone hole, 23

P

Paclitaxel
  historical perspective, 52
  isomers, 54–56
  medical value of, 53
  modern-day preparation of, 53
  nature of, 52
  nuclear magnetic resonance (NMR) spectroscopy, 56–58
  terpenes and isoprene, 53–54
Palmitic acid, 62, 111, 123, 178
Palm oil, 178
Paracelsus, 133
Paracetamol, 153–154
Parietin, 220, 223–224
Parmelia saxatilis, *see* Crottal
Pasteur, Louis, 93
p-benzoquinone, 217
Pelletier, Pierre, 32
Pennyroyal, 14

Peptide bond, 68, 73–74
Peptide link, 7, 66
Periwinkle, 46–47
  alkaloids and, 49–50
  research and therapeutic value, 48–49
  serendipity, 47
  vinblastine, 47, 48, 49–50
  vincristine, 47, 48–49, 50
Perkin, William Henry, 3, 33, 212, 216–217
Perm, 98
Peroxides, 31, 35–36
Pessaries, 13–14
Pesticides, 79
Petroleum, 20–21, 99, 136
Pharmacognosy, 2–3
Phenolphthalein, 233
Phenols, 27–28, 107–108, 200, 218
  compared with alcohols, 107
  definition, 107
  electrophilic substitution reactions of, 108
Phenoxazones, 232
Phenylalanine, 68, 75, 76, 78
Phenyl ethanoate, 28
Phenylpropanoids, 66, 68, 69, 78
Photochemical smog, 25, 37
Photoperiodism, 233–234
Photosynthesis, 82, 94–95, 191, 193, 201
Phthalo blue, 226
Phytochrome, 228, 233–234
Pi bonds, 196, 197
Pincus, Gregory, 17
Piperidine, 147
Plantae, 191
Plant kingdom, 5, 189, 191–192
Pliny the Elder, 212
Pollution, 23, 99, 161–163, 200, 224
Polymers, 7, 69, 73, 74, 143, 181
Polyphenols, 107, 113
Polysaccharides, 83, 85–86, 87, 181
Poppy seeds, 133, 134
Porphyrin, 193
Posselt, Wilhelm Heinrich, 151
Priestley, Joseph, 193
Progesterone, 13, 14, 18
  biological studies of, 17
  chemical structure of, 14
  to prevent ovulation, 17
Proline, 75
Protein, 56, 66, 68, 72–73, 74–76, 97, 153
  amino acids and, 74–75
  importance of, 72–73
  molecular structure of, NMR and, 75–76
Protista, 191
Pyridine, 151, 152–153, 154
Pyrrole, 151, 152–153, 154, 193, 233
Pyruvate, 93–94

Q

Queen Anne's Lace, *see Daucus carota*
Quinine, 31, 32–34, 146, 216
  isolation of, 32–33
  synthesis of, 33
  uses of, 33–34
Quinoa, 61
Quinones, 220–221

R

Racemic mixtures, 160
Radiation
  infrared, 161–163, 200, 201, 233, 234
  ultraviolet, 23, 199, 220, 221
  visible, 198
Rearrangement reactions, 142, 170
Redox reaction, 90, 93–94, 98
Reduction, 153, 200, 201, 202, 215
Reimann, Karl Ludwig, 151
Resveratrol, 234
Retinal, 210–211
Retinol, 210
Reversible colors; *see also* Reversible dyes
  anthocyanins, 231
  autumn leaves, 229
  flavonoids, 229–231
  inestimable value of color, 228–229
  phytochrome, 233–234
  reversible dyes, 231–233
Rhodopsin, 210
Ribose, 87
Rice, 82
Rocella, see Royal Purple
Roccella tinctoria, 231
Roselius, Ludwig, 119
Rosenkranz, George, 16
Rosmarinic acid, 78
Royal Botanic Gardens, Kew,
    United Kingdom, 163
Royal Purple, 223
Rubia tinctorum, 223

S

Saccharides, 82, 83
  disaccharides, 82, 83–84
  monosaccharides, 82, 83, 85, 86–87
  oligosaccharides, 82, 85
  polysaccharides, 82, 83, 85–86, 87, 181
Saffron, 204, 205
Salicin, 26
Salicylic acid, 26, 28–29
  acetylation of, to form aspirin, 28–29
  molecular structure of, 26

Salix alba, 29
Sallaki Guggul, *see* Frankincense
Salts, 147, 154, 178–179
Salvia hispanica, 60
Sanger, Margaret, 17
Sapogenins, 15
Saponification, 173, 177
Saturated compounds, 19, 63, 199, 209
Saturated hydrocarbons, 19–20
Scurvy, 43
Serine, 75
Sertürner, Wilhelm, 135
Shen Nung, 102
Shikimic acid pathway, 66, 76–78
Silphium, 14
Skin treatment and *Aloe vera*, 181
Sloane, Hans, 112
Soap, 175, 177–179
Sodium stearate, 178
Solar radiation, 162
Squalene, 53, 142
S–S bonds, 97, 98
Starch, 71–72, 85, 87–88, 104
Steam distillation, 159, 160
Stearic acid, 62, 111, 178
Stereochemistry, isomers and, 54–55
Steric hindrance, 171, 213
Steric protection, 171
Steroid, 13, 15, 38, 53, 170
  esters from alcohols and acids, 40–42
  foxglove and digoxigenin, 38–40
  lactones, 42, 44
  vitamins, 42–44
Steroid saponins, 15
Strychnine, 4, 126–127
Sucrose, 82, 83–85
Sugars, 83, 86, 93, 181, 182, 206
Sulfur-containing compounds, 97
Supercritical carbon dioxide, 119–120
Surfactant, 178
Svoboda, Gordon H., 48, 49

T

Tansy, 216
Tartrazine, 218–219
Taxol®, *see* paclitaxel
Taxus baccata, 53
T. dicoccum, 71
Tea, 101; *see also* Beverages
  art of processing, 103
  in China, 103
  in England, 104
  extraction processes, 104–105
  growing and processing of, 104–105
  history of, 102
  as modern medicine, 103, 105–106
  opium and, 103
  spread of, throughout world, 103–104
Terpenes, 52, 53–54, 141–143, 158, 168–170
  boswellic acid, 167, 168, 170
  classification of, 143
  cyclic, 142, 170
  definition, 141
  elimination reactions and, 142–143
  isoprene rule, 140, 142, 170
Terpenoids, 52, 53, 158, 206, 208
  camphor, 54, 158–160, 168, 216
  cyclic,142, 158, 165, 168–171, 216
  menthol, 54, 158
δ 1-tetrahydrocannabinol (THC), 140–141
Tetramethylsilane, 58
Textile dyes
  colorfastness, 225
  mordants, 225–226
  solubility in water, 225
Theobroma cacao, 110, 113
Theobromine, 114
Thin-layer and column chromatography, 185
Thiols, 99
Thiophen, 24, 152–153
Threonine, 75
Thujone, 216
T. monococcum, 71
Tobacco, 149, 150
  amides, 153–154
  amines, 151
  heterocyclic aromatic compounds, 151–153
  nicotine, 149, 151, 154–155
Traditional medicine, 2, 11, 184, 222
Trans-esterification, 63–64
Trichlorophenol, 108
Triglycerides, 63, 64, 177
Triglyceryl esters, 63
Triticum, 66, 67
Tropane, 146–147
Troposphere, 25
Tryptophan, 75, 76, 124
Tubocurarine, 3, 128, 129
Twining, Thomas, 104
Tyrosine, 75, 76, 77, 78

U

UK Climate Change Act, 202
Ullmann's Encyclopedia of Industrial Chemistry, 215
Ultraviolet absorption spectroscopy, 198
  molecular structure and, 199–200
  organic chemistry and, 198–199
Ultraviolet radiation, 23, 199, 221
Unsaturated compounds, 62, 63, 199, 208, 232

V

Valine, 75
Van der Waals forces, 72
Vegetable oil, 63, 64–65, 176–177
Vinblastine, 46, 47, 48, 49–50
Vincristine, 47, 48–49, 50, 146
Vinculin, 47
Vinyl, 209
Visible radiation, 198
Vitamin A, 43, 204, 207, 210–211
Vitamin C, 39, 42, 43–44
Vitamin K, 43
Vitamins, 5, 42–43, 72; *see also individual vitamins*
von Baeyer, Adolf, 212, 217

W

Wald, George, 210
Warm-season cereals, 70
Water-soluble vitamins, 43
Waxes, 64
Wheat
- condensation reaction, 73–74
- crop domestication, 67
- crop germination, 67
- eugenol and rosmarinic acid, 78
- genetic modification of crops, 78–80
- genetics, 71
- lignans, 68–69
- lignin, 69
- major cultivated species of, 69–70
- modern, 70–71
- NMR and proteins, 75–76
- nutritional importance of, 71–72
- origins of ancient, 66–67
- peptide bond, 73
- phenylpropanoids, 68
- proteins and amino acids, 72–75
- shikimic acid pathway, 76–78

White indigo, 197, 215
Whittaker, Robert, 191
Willow bark, 26, 29
Withering, William, 39
Woad, *see Isatis tinctoria*
Wrangler, 215

X

Xanthophylls, 206, 207
- crocin, 208
- lutein, 208–209

Xanthoria parietina, 223–224
Xocolatyl, 110–111

Y

Yeast, 93

Z

Zwitterions, 118
Zymology, 93

PGMO 04/26/2018